SOCIETY FOR THE STUDY OF HUMAN BIOLOGY

SYMPOSIUM SERIES: 27

Genetic variation and its maintenance

PUBLISHED SYMPOSIA OF THE

SOCIETY FOR THE STUDY OF HUMAN BIOLOGY

1 The Scope of Physical Anthropology and its Place in Academic Studies
Ed. D. F. ROBERTS and J. S. WEINER
2 Natural Selection in Human Populations Ed. D. F. ROBERTS and G. A. HARRISON
3 Human Growth Ed. J. M. TANNER
4 Genetical Variation in Human Populations Ed. G. A. HARRISON
5 Dental Anthropology Ed. D. R. BROTHWELL
6 Teaching and Research in Human Biology Ed. G. A. HARRISON
7 Human Body Composition, Approaches and Applications Ed. J. BROZEK
8 Skeletal Biology of Earlier Human Populations Ed. D. R. BROTHWELL
9 Human Ecology in the Tropics Ed. J. P. GARLICK and R. W. J. KEAY
10 Biological Aspects of Demography Ed. W. BRASS
11 Human Evolution Ed. M. H. DAY
12 Genetic Variation in Britain Ed. D. F. ROBERTS and E. SUNDERLAND
13 Human Variation and Natural Selection Ed. D. F. ROBERTS
(Penrose Memorial Volume reprint)
14 Chromosome Variation in Human Evolution Ed. A. J. BOYCE
15 Biology of Human Foetal Growth Ed. D. F. ROBERTS
16 Human Ecology in the Tropics Ed. J. P. GARLICK and R. W. J. DAY
17 Physiological Variation and its Genetic Basis Ed. J. S. WEINER
18 Human Behaviour and Adaptation Ed. N. G. BLURTON JONES and V. REYNOLDS
19 Demographic Patterns in Developed Societies Ed. R. W. HIORNS
20 Disease and Urbanisation Ed. E. J. CLEGG and J. P. GARLICK
21 Aspects of Human Evolution Ed. C. B. STRINGER
22 Energy and Effort Ed. G. A. HARRISON
23 Migration and Mobility Ed. A. J. BOYCE
24 Sexual Dimorphism Ed. F. NEWCOMBE *et al.*
25 The Biology of Human Ageing Ed. A. H. BITTLES and K. J. COLLINS
26 Capacity for Work in the Tropics Ed. K. J. COLLINS and D. F. ROBERTS
27 Genetic Variation and its Maintenance Ed. D. F. ROBERTS and G. F. DE STEFANO

Numbers 1–9 were published by Pergamon Press, Headington Hill Hall, Headington, Oxford OX3 0BY. Numbers 10–24 were published by Taylor & Francis Ltd, 10–14 Macklin Street, London WC2B 5NF. Further details and prices of back-list numbers are available from the Secretary of the Society for the Study of Human Biology.

Genetic variation and its maintenance

with particular reference to tropical populations

Edited by

D. F. ROBERTS

Department of Human Genetics, University of Newcastle upon Tyne

G. F. DE STEFANO

Dipartimento di Biologia, II Universita di Roma (Tor Vergata)

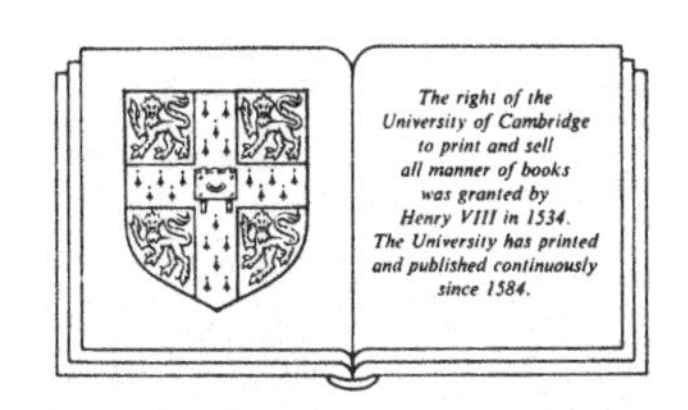

CAMBRIDGE UNIVERSITY PRESS

Cambridge

London New York New Rochelle

Melbourne Sydney

CAMBRIDGE UNIVERSITY PRESS
Cambridge, New York, Melbourne, Madrid, Cape Town, Singapore, São Paulo

Cambridge University Press
The Edinburgh Building, Cambridge CB2 8RU, UK

Published in the United States of America by Cambridge University Press, New York

www.cambridge.org
Information on this title: www.cambridge.org/9780521332576

First published 1986
This digitally printed version 2008

A catalogue record for this publication is available from the British Library

ISBN 978-0-521-33257-6 hardback
ISBN 978-0-521-06457-6 paperback

CONTENTS

PREFACE

The fact that human populations differ in their normal gene frequencies was first shown by the Hirszfelds in 1918 in their survey of the ABO bloodgroups in soldiers and others of different nationalities. The Mendelian basis of the ABO blood groups was already known in 1910, though several years were to pass before the exact mechanism of their inheritance was established by Bernstein in 1924. Included in the Hirszfelds' material were the first samples of tropical populations, e.g. from Cambodia, Vietnam, French West Africa, and the central provinces of India. Over the years, knowledge of gene frequencies in tropical populations increased gradually, but then it received considerable impetus some two decades ago with the development of the International Biological Program. Indeed, much of the present information on human genetic variation in the tropics derives from investigations carried out under its auspices. Information on the genetic constitution of populations was collected with the intention of using it as a variable to be controlled in comparisons of groups inhabiting different environments, and also in its own right with a view to understanding the affinities and differentiation of the communities investigated. At that time, however, although most of the polymorphic blood group systems had been established, the usefulness of electrophoresis as a technique for distinguishing genetic variants was only just beginning to be understood and applied. Hence from the surveys of that period genetic information is restricted to gene frequencies in most polymorphic blood group systems, and a few serum and red cell enzyme and other protein systems.

Since that time, however, knowledge of genetic variablity in man has increased enormously. Older laboratory techniques have been refined and new ones devised for examination of these established systems (e.g. two-dimensional electrophoresis, isoelectric focusing, use of monoclonal antibodies), so that what were formerly thought to be single Mendelian variants have been found to consist of a number of subtypes, as for example with red cell phosphoglucomutase or serum vitamin D binding protein. Entirely new polymorphic systems have been discovered and exploited, and particularly those of many enzyme systems and the major histocompatiblity complex. At one extreme, microscopically visible variation in the detailed morphology of the human chromosomes has been revealed by differential staining procedures. At the other, full aminoacid and base pair sequencing of alleles has been carried out in a number of polymorphisms, and the molecular differences between variants established, so that again what appeared at the first level of analysis to be a single entity proved to conceal considerable heterogeneity. The thalassemias provided an illustration of such heterogeneity. Here, with the establishment of the fundamental structure of haemoglobin, first there came the differentiation of those affecting the alpha and beta chains of the molecule. The thalassemias were recognised as being due to the reduced or absent synthesis of the affected globin chain, in the presence of the continuing synthesis of the unaffected chain. But then sequence analyses demonstrated a variety of mutations. Some affected the chain terminator of the coding region, others consisted of deletions or changes of single or multiple nucleotides; in general, deletions were found to account for most of the alpha thalassemias, and point mutations producing defects in transcription or in mRNA processing to account for most of the beta variants. Today, some 40 thalassemia variants (9 α and 31 β) are known, and they tend to be population specific; of the β mutations only two are present in more than one continental group.

Moreover, sequence analysis has been extended to the DNA itself, showing the existence of intervening sequences between the codons within the coding sequence, of flanking sequences, of pseudogenes. Development of restriction endonucleases and their use to identify the presence or absence of particular sequences and so to give restriction fragment length polymorphisms has revealed another enormous source of variablity within these flanking sequences. These have begun to be applied in clinical genetics for presymptomatic and prenatal diagnosis, and their limited application at the population level already promises that they will be of enormous utility. The first hint of this was in 1980, when use of a single restriction enzyme (Hpa 1) showed that the gene responsible for the beta chain of HbS was different in West and

East Africans. Now, for example, in the beta globin chain cluster, some 60 kb in length, eleven polymorphic restriction sites, identified by eight restriction enzymes, give three distinct haplotypes associated with the β^S gene. In Africa these are geographically specific, associated with Benin, Senegal, and Central Africa. Clearly these extremely powerful polymorphisms are of relevance in demonstrating local and regional variation, but they also indicate that discrete mutations occurred at different geographical localities and therefore they show something of the origin and movement of the populations in whom they are observed. Moreover, they suggest that older estimates of mutation rates are likely to be inadequate, and so are all the estimates of genetic variability and the uses to which they have been put in quantification of evolutionary rates, genetic distances between populations, genetic relationship, and similarly. These discoveries mean that the current descriptions of genetic constitution of tropical populations are merely skeletons. The time is now ripe to put flesh on these, and to give the additional dimensions that will allow fuller interpretation and understanding of human genetic variability in the tropics.

A working party was therefore convened, sponsored by the International Union of Biological Sciences, in Frascati, Rome, in April 1985. As a preliminary to its discussions, a symposium to remind participants of what has been achieved and what may be expected was organised on Genetic Variation and its Maintenance in Tropical Human Populations. A selection of the papers presented at that meeting, intended to be illustrative rather than comprehensive, is published in the present volume.

The first section deals with the dimensions of genetic variation. It concerns characters about the transmission of which there is no doubt. Not only is their genetic basis known, but also in many instances their biochemistry, their molecular configuration, and their structure down to the very detail of the aminoacid sequences, the DNA base pair sequences that code for them, and the noncoding flanking sequences in which they are framed, and variations in the codons that have produced their variants. The section shows how more sophisticated techniques applied to systems already known refine the information that is provided, and alter the interpretation of the data; how differing frequencies of phenotypes of polymorphisms discovered since the International Biological Program characterise tropical populations; how little is known of chromosomal polymorphisms in populations at the level of microscopically detectable variation; and by contrast, how much may be expected from the enormous potential of the molecular polymorphisms of the DNA in the nucleus and mitochondria.

The second part concerns the origin and maintenance of genetic variability, as illustrated principally by the more simply inherited characters covered in section 1. Data from the Pacific and from African populations are reviewed and the selective challenge to man of disease and his response is illustrated with reference to malaria. The potential of the recently discovered haplotype heterogeneity of the Hb^S gene emerges clearly. How selection may be further explored by laboratory enquiry is illustrated in relation to the albumin locus. The importance of random change in a bottleneck situation and subsequent hybridisation in the production of new gene pools is illustrated with reference to Central America, and migration allowing such hybridisation illustrated in Central Africa. A frequently overlooked factor in the maintenance of genetic variability, the genetic structure of the population and its mating patterns, is reviewed.

The papers in the final section of the book intentionally contrast with those in the first two, for they concern characters in which the genetic contribution to variability is complex. The section serves as a reminder that genetic variability of human populations is not only manifested by single genes but also by polygenic complexes which, interacting with the developmental milieu, determine the phenotype. There are many such features, in the development of which there is a degree of genetic control that ranges from slight to maximal. Among these features occur the majority of physiological variables, so important in the fine tuning of man's adjustment to the vagaries of his environment; the majority of the outward and visible features that may be involved in choice of spouse; the majority of performance variables, mental and physical, and many others. Where such features have a high degree of genetic control they may be usefully employed to provide information on population affinities and distances that complement those emanating from the monogenic characters. Those that are less strictly controlled genetically can be used to trace similarities in response to common environmental stresses in genetically differing populations, and so help to resolve problems of adaptation. The section shows how such characters may be used to elucidate biological problems of affinity and differentiation, of adaptation and survival. It raises the question of the importance of phenotypic plasticity in helping maintain genetic diversity, and it gives examples of how the genetic peculiarities of small isolated populations may be successfully exploited to explore the etiology of diseases.

The selection in the volume as a whole points out a number of problems awaiting attention, shows the potential of the new techniques and the types of problem for which they are particularly suited, and the results of some of the applications.

Following the symposium the members of the Working Party were able to draw up a series of recommendations for research under the auspices of the IUBS Decade of the Tropics programme, which will be published in Biology International in 1986. It is hoped that the papers published in the present volume will be as informative and stimulating to a wider readership as they were to the symposium participants.

D. F. ROBERTS
Department of Human Genetics
University of Newcastle upon Tyne, England

G. F. DE STEFANO
Dipartimento di Biologia
Seconda Universita di Roma (Tor Vergata)
Rome, Italy

PART I

GENETIC DIVERSITY - ITS DIMENSIONS

GENETIC POLYMORPHISMS - A WIDENING PANORAMA

D. F. ROBERTS

Department of Human Genetics,
University of Newcastle upon Tyne, England

INTRODUCTION

Sir Christopher Wren in 1657 was the first to make intravenous therapy possible, in that he devised an instrument - a needle made from a slender quill fixed to a bladder - by which substances could be injected. As a result, over the next two and a half centuries the sequelae that often followed transfusions came to be recognised, and were documented especially during the Franco-German war when many attempts were made to help the wounded by transfusion under field conditions. But it was the discovery (1875) of serological species specificity, in that mixture of red blood cells of an animal of one species with serum from one of another *in vitro* leads to agglutination of the red cells, that triggered the discovery of the first blood polymorphism in man. For it was this that prompted Landsteiner to enquire whether differences in agglutination, similar to those in interspecific mixtures, occurred between individuals of the same species.

SYSTEMS

Red cell blood groups

The simplest method of investigation, mixing the serum of one person with the red blood cells of another, led Landsteiner to discover what subsequently became known as the A, B and O blood groups in 1900. The fourth (AB) group was discovered two years later. Then at first there was little advance. In 1910 the Mendelian inheritance of the ABO system was established, and in the same year the A1-A2 subdivisions were discovered by von Dungern and Hirszfeld. That the frequencies differed from one population to another was established by the Hirszfelds' examination of soldiers and prisoners of war of different nationalities. But it was not until 1927 that Landsteiner and Levine published the results of their deliberate search for antibodies which would classify blood into groups independently of the known ABO distinctions. They announced the discovery of M, N, and P in 1927, following experiments in which rabbits which had been immunised with human red cells were found to produce sera which, after absorption of the species agglutinins, would agglutinate some specimens of blood but not others.

These early papers provided the method by which many subsequent systems were established. First comes the identification of the antigen, then differences in frequencies are examined in different populations, while family studies show the inherited basis. Discoveries followed slowly at first, then at an increasingly rapid rate. In 1932 was discovered secretor, 1935 H, 1936 A3, 1938 M2, and then in 1939 a major discovery, that a blood group was responsible for the severe haemolytic disease of the newborn. Levine and Stetson described how the mother of a stillborn fetus suffered a severe haemolytic reaction to transfusion from her husband, and on further examination she was found to have an antibody that agglutinated the cells of 80 out of 104 ABO compatible donors. The antigen responsible was independent of the ABO, MN and P groups, and an antibody could not be made by immunisation of rabbits. Their interpretation was that the mother had been immunised by her fetus, which had inherited from its father an antigen which the mother lacked, so that the mother reacted to the fetal cells in the same way as she did to her husband's blood. The discovery followed that rabbit immunisation against the blood of the rhesus monkey produced antibodies that agglutinated the red cells of a large proportion of white people in New York, who were therefore called rhesus positive; and that these antibodies occurred in the serum of people who had shown transfusion reactions despite ABO compatibility with the donor. The remaining link was to show that this antibody was the same as that found in Levine and Stetson's subject. It took several decades more for it to be realised that rabbit anti-rhesus and the human anti-Rh antibodies are not identical.

There followed the rapid discovery of the further C, c, and E alleles of the rhesus system by 1943, when Fisher saw that the results of tests using the four corresponding antisera included reactions that were antithetical, and argued the presence of three pairs of genes at very closely linked loci. This led to the discovery of previously unknown reactions which occurred as predicted by Fisher's theory, and then in 1945 anti-e was discovered by Mourant, and all the expected interactions of Fisher's theory were observed. The few years that followed were indeed the hey-day of blood group discoveries, for the other major systems came to be identified very rapidly, until by the early '60s all the major blood group systems known today on the surface of the red cells had been discovered. Later work consisted of their more detailed investigation, e.g. of the chromosomal locations of the genes responsible, and the biochemical constitution of these complex macromolecules.

The pattern of discovery (Fig. 1) is therefore rather an asymmetrical S-shaped curve - a slow start, then accelerating progress, then deceleration. Few systems have been the result of deliberate search, and most antibodies have been discovered by intensive investigation of the serum of patients who had received numerous blood transfusions, or in women who had produced children with haemolytic disease of the newborn.

Gm groups

While the rate of discovery of bloodgroup polymorphisms was at its peak, a new group of variants was encountered, initially by chance, and established using the techniques of inhibition of agglutination and, later, precipitation reactions and immunodiffusion. These were the Gm and related systems.

The Gm allotypes have proved to be highly polymorphic, and indeed are the second most polymorphic system known in man. The existence of the Gm system was first discovered by Grubb (1956) who was examining the immunoglobulin G (IgG) levels in the serum of patients with rheumatoid arthritis. He observed that Rh+ red cells coated with incomplete IgG and anti-D antibodies were agglutinated by the serum of one patient. He found that this agglutinating ability was relatively common in serum from patients with rheumatoid arthritis, and that serum from about 60% of normal subjects would inhibit this agglutination. He and Laurell (1956) showed that this blocking ability was genetically determined; it was inherited as a simple autosomal dominant. Because it was found in the gammaglobulin, the factor was called Gm, the inhibiting sera being Gm(a+) and the non-inhibiting Gm(a-).

There followed rapidly the discovery of additional antigens in the Gm system. Gm^{2} was discovered in 1959 as a result of finding anti-Gm^{2} in a patient with rheumatoid arthritis, and so was the original anti-Gm^{7} (1961), anti-Gm^{8} (1962), anti-Gm^{9} (1963), anti-Gm^{15} (1966); others were discovered in normal subjects, Gm^{10} (1963), Gm^{11} (1963), Gm^{13}, Gm^{14}, while yet other antisera were discovered by deliberate experimentation, anti-Gm^{17} & ^{22} deriving from immunised rabbits. In 1960, using a method similar to that for Gm determination, Ropartz and colleagues reported a system (Inv) which stopped the agglutination of sensitized O Rh+ red cells, but was independent of Gm. Other alleles were soon identified, with the discovery of anti-Inv 2 (1961) and Inv 3 (1962). A2M was discovered independently by two teams in 1969, one of them finding the

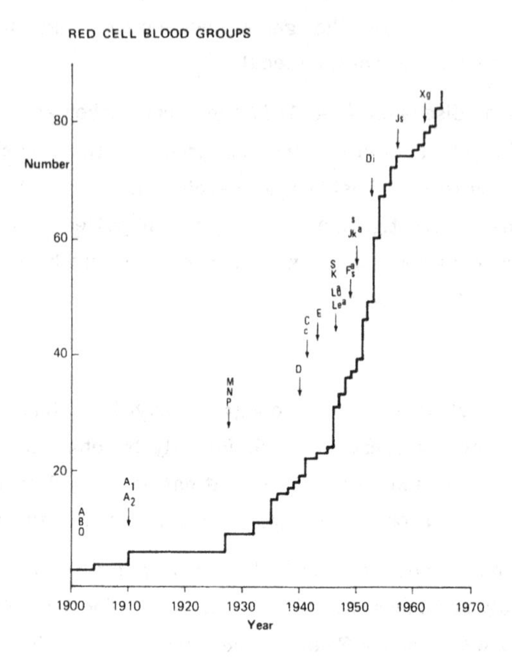

Figure 1: **Discovery of the major polymorphisms: (i) red cell blood groups.**

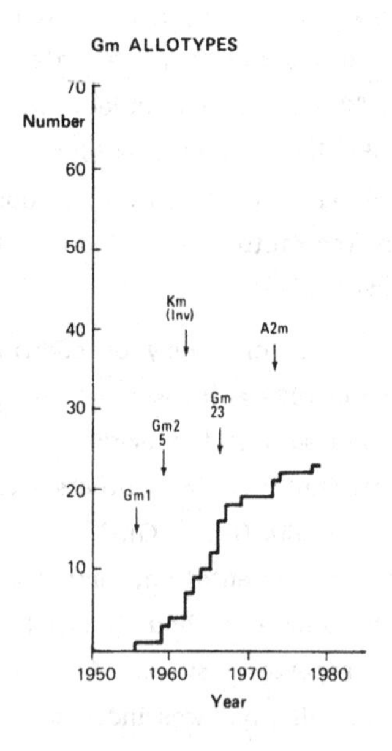

Figure 2: **Discovery of the major polymorphisms: (ii) Gm allotypes**

antibody in a subject who had experienced a transfusion reaction. Like Inv it is detected by blocking agglutination, but this time the indicator cells are coupled with a myeloma protein, since there are no incomplete anti-red cell antigen antibodies that are IgA 2.

The discovery of the location of the Gm groups was exciting. Not only are they located within IgG, but different Gm antigens occur in different heavy-chain subclasses. Thus, Gm 1, 2 and 17 are carried by γ1, Gm 23 by γ2, and Gm 5, 6 and 10 by γ3. The Inv antigens, however, are carried by the light (kappa) chains, and hence the change of name to Km allotypes instead of Inv. Like Gm, Am is carried in the heavy immunoglobulin chain, but in that of IgA 2 instead of that of IgG. The Gm markers on the different subclasses of IgG, and Am markers on IgA, often differ only in single amino-acid substitutions. It is the location of the subgroup that accounts for the fact that anti-D antisera can be used for detection of the γ1 and γ3 groups, but not Gm 23 on the γ2; anti-D antibodies are almost invariably γ1 or γ3 or both. The rate of discovery of new variants (Fig. 2) was not as rapid as that of the red cell blood groups at its peak and has now slowed.

Biochemical polymorphisms

By the 1960s the impetus attaching to blood group discoveries had shifted to the biochemical polymorphisms. Pauling's demonstration in 1949 that the major haemoglobin fraction in blood from sickle cell anaemia patients differed in electrophoretic mobility from that in normals, and that heterozygotes could be similarly differentiated by the presence of two protein bands instead of one, set the stage for many subsequent applications of the electrophoretic technique. Major advance followed its refinement by Smithies in 1955 using starch gel instead of paper, which gave much greater resolving power; this derived from the more sensitive molecular sieving obtained by the similarity in distributions of the pore sizes of the starch gel and the molecular sizes of the proteins. Starch gel was joined by acrylamide, agarose and other gels as supporting media, which all have relative advantages and disadvantages for separation of particular protein systems, and further refinements came with the development of rocket electrophoresis, immunoelectrophoresis, two-dimensional electrophoresis, and other procedures. As a result there was a veritable blossoming in the discoveries of biochemical polymorphisms (Fig. 3).

The first polymorphism to be identified was that of haemoglobin S, and the existence of haemoglobin C in the following year, but then Smithies'

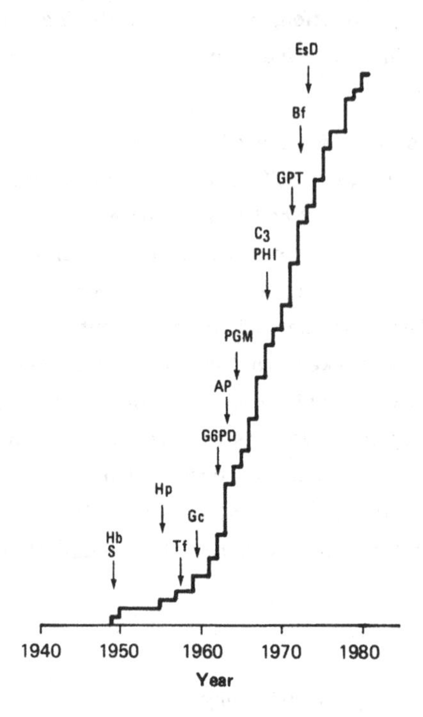

Figure 3: Discovery of the major polymorphisms: (iii) biochemical.

discovery of the haptoglobin types initiated a phase of rapid advance. The serum proteins transferrin and group-specific component came first, identified respectively by starch gel electrophoresis and immunoelectrophoresis, and then the red cell enzymes joined them - red cell acid phosphatase (1963), glucose-6-phosphate dehydrogenase (1962), 6-phosphogluconate dehydrogenase (1963), phosphoglucomutase (1964), adenylate kinase (1966) and many more. These polymorphisms were all the result of deliberate search. Known enzymes thought to be homogeneous were attacked by batteries of relevant substances over a wide range of experimental conditions and of visualising media, until isoenzymes were discovered. The result was a protracted series of discoveries sustained over twenty years (1960-80) at a rate that matched the maximum in blood group discoveries over the ten-year period 1945-1955. Only now is there the suspicion that the rate of discovery may be beginning to fall off.

Polymorphisms with a biochemical basis occur in many other tissues (brain, liver, muscle, heart, etc.) and body substances, but few studies of population differences in these have been made. There are a few notable exceptions, for example in pharmacogenetics the rates of metabolism of isoniazid and of beta-amino isobutyric acid; and in perception, where the biochemical basis is less clear, in the ability to taste phenylthiocarbamide (one of the earliest polymorphisms to be discovered) and colour vision. In this group as a whole, however, polymorphisms are too few to attempt an analysis of discovery rate, except in one group of polymorphisms, the salivary enzymes.

Following the discovery of genetically-determined salivary amylases in the mouse, amylase isoenzymes were soon found in man in 1965, their genetic basis established, and their gene frequencies in some populations. There followed in quick succession discovery of the salivary vitamin B12 binding protein, the parotid (1972) and post-parotid basic proteins (1977), the parotid proline-rich protein (1974), and the parotid double band protein. Today there are at least nine loci that determine parotid salivary proteins and the products of at least seven genes comprise the parotid-rich protein complex. The acidic and basic proline-rich proteins constitute about two-thirds of the parotid salivary proteins. Salivary peroxidase polymorphism was discovered in 1977, and the salivary acid phosphatase A in 1976, and in the same year the salivary esterase proteins. These are relatively recent additions to the armamentarium of human polymorphisms, and it seems likely that many others remain to be discovered. They are likely to be of particular interest in terms of selective differentials. The rate of discovery seems to be still in its middle rapid phase.

Polymorphisms of the major histocompatibility complex

Our present knowledge of the polymorphisms at loci in the major histocompatibility complex is the result of some 30 years' work in man, again like the blood groups based on earlier observations in animals. Gorer (1936) reported that graft rejection in rodents was genetically based, and Medawar's work showed that this rejection was antigen-specific and immunological (1944). Snell (1948) introduced the concept of histocompatibility genes, and assigned the title H2 to this system in the mouse, and analogous systems were subsequently identified in all vertebrates (Gotze, 1977). Application to man had to await appropriate techniques, but Dausset (1954) described an antigen (Mac) located on the surface of human blood leukocytes and recognised by antibodies in the serum of polytransfused individuals. Similar antibodies were then reported in multiparous women. By 1964 a sensitive microlymphocytotoxicity test for the identification of antigens had been introduced, two groups of antigens had been established (van Rood & van Leunen, 1963; Payne et al, 1964), and the exploration of the human MHC system was under way.

In the MHC region on the short arm of chromosome 6, are situated the genes for the HLA antigens, which are inherited as codominant traits. There are several loci and it took some time to distinguish their precise relationship. The gene products are today divided into three classes. Class I (three loci) consists of HLA-A, B and C products that occur on the surface of all nucleated cells except sperm and trophoblasts. These antigens have two polypeptide chains. The alpha-chain, where the specificity resides, maps to the MHC region, while the second chain, composed of beta-2-microglobulin, maps to chromosome 15. Class II with at least three loci includes D, DP, DQ and DR gene products. These are restricted in distribution, having their highest concentration on B cells. Class III includes the proteins in the complement system (Bf, C2, C4a and C4b).

In the major histocompatibility complex are located the most polymorphic systems yet discovered in man. No single allele is very common, and no other system shows such high levels of heterozygosity. There is considerable variation in antigen frequencies between the different racial groups, both for the Class I and Class II antigens. Also, there is considerable variation among human populations in the linkage disequilibrium that occurs. But even now the extent of polymorphism in the MHC region has not been finally defined, for it seems likely that, for example, the Ir gene products (immune response genes) that have been found in animals, will also be found in man, or at least others equivalent to them, and will also prove to be polymorphic. Examples of such genes related to Ir that have been found in man are those which induce suppression (Is genes) (Sasazuki et al, 1980; Hayes et al, 1982; Solinger et al, 1982).

The rate of discovery is extremely difficult to plot on account of multiple identifications of the same or related antigens, reconciliation and subdivision. Suffice it that the rate of antigen identification exceeds considerably that of variants in the other types of polymorphism. This is due to the remarkable organisation attending the investigation of this system, with intense international collaboration, regular organised workshops, and worldwide exchange of antisera between laboratories. To this, the role of these antigens in relation to the autoimmune disorders and in tissue transplantation was an undoubted stimulus.

DISCUSSION

Progress in knowledge of the human polymorphisms therefore resembles a relay race, in that there has been a succession of participants, each taking over when its predecessor appears to have been losing impetus. The red cell blood groups led the field up till the early 1960s, giving place to the biochemical systems from 1965 onwards, side by side with the immunoglobulin allotypes, which in turn were succeeded by the salivary proteins from 1970 onwards, and the remarkable development of the polymorphisms of the major histocompatibility complex. Of these, the shape of the curve of discovery indicates that there is likely to be more still to be done in the biochemical systems and in the salivary proteins, and at other loci linked to the HLA.

That is the present state. The last 2-3 years have seen the commencement of exploitation of a number of new techniques, identifying new polymorphisms. These do not concern gene products as do all those so far mentioned. Some are at the grosser level, chromosome segments. But perhaps the most far-reaching are those that descend below the gene level to the DNA itself and relate to variations in base sequence. Here we are at the beginning of another S-shaped curve, which already promises to be more exciting, more informative, more detailed, and of greater theoretical utility in the study of man and practical value in management of his genetic morbidity than all of its predecessors combined. It is difficult for our imagination to grasp the infinity of detail that the next few years will bring.

REFERENCES

Dausset, J. (1954). Leuko-agglutinins, IV: Leukoagglutinins and blood transfusion. Vox Sanguinis, **4**, 190.

Gorer, P.A. (1936). The detection of antigenic differences in mouse erythrocytes by the employment of immune sera. British Journal of Experimental Pathology, **17**, 42.

Gotze, D. (ed.) (1977). The Major Histocompatibility System in Man and Animals. Berlin: Springer.

Grubb, R. (1956). Agglutination of erythrocytes coated with "incomplete" anti-Rh by certain rheumatoid arthritic sera and some other sera. Acta Pathologica Microbiologica Scandinavica, **39**, 195.

Grubb, R. & Laurell, A. B. (1956). Hereditary serological human serum groups. Acta Pathological Microbiologica Scandinavica, **39**, 390.

Hays, E. F., Jones, P., Fathman, C. G. & Engelman, E. G. (1982). Response to streptococcal cell wall antigens. Clinical Immunology and Immunopathology, **25**, 283.

Landsteiner, K. (1900). Zur Kenntnis der antifermentativen, lytischen und agglutinierenden Wirkungen des Blutserums und der Lymphe. Zeitblat Bakteriology, **27**, 357.

Landsteiner, K. & Levine, P. (1927). A new agglutinable factor differentiating individual human bloods. Proceedings of the Society for Experimental Biology, New York, **24**, 600.

Medawar, P.B. (1944). The behaviour and fate of skin autografts and skin homographs in rabbits. Journal of Anatomy, **78**, 176.

Pauling, L., Itano, H. A., Singer, S. J. & Wells, I. C. (1949). Sickle cell anemia, a molecular disease. Science, **110**, 543.

Payne, R., Tripp, M., Weigle, J., Bodmer, W. & Bodmer, J. (1964). A new leukocyte iso-antigen system in man. Cold Spring Harbour Quantitative Biology, **29**, 285.

Ropartz, C., Lenoir, J., Hemet, Y. & Rivat, L. (1960). Possible origins of the anti-Gm sera. Nature, **188**, 1120.

Sasazuki, T., Kanoka, H., Nishomura, Y., Kaneoka, R., Hayama, M. & Ohkuni, H. (1980). An HLA-linked immune suppression gene in man. Journal of Experimental Medicine, **152**, 297.

Smithies, O. (1955). Zone electrophoresis in starch gels: Group variation in the serum proteins of normal human adults. Biochemical Journal, **61**, 629.

Snell, G. D. (1948). Methods for the study of histocompatibility genes. Journal of Genetics, **49**, 87.

Solinger, A., Bhatnager, R. S. & Stobo, J. D. (1982). Cellular molecular and genetic characteristics of T cell reactivity to collagen in man. Proceedings of the National Academy of Sciences, **78**, 3877.

Van Rood, J. J. & Van Leeuwen, A. (1963). Leukocyte grouping: a method and its application. Journal of Clinical Investigation, **42**, 1382.

SOME IMPLICATIONS OF IMPROVED ELECTROPHORESIS TECHNIQUES FOR POPULATION GENETICS

S. S. PAPIHA

Department of Human Genetics, University of Newcastle upon Tyne, Newcastle upon Tyne, U.K.

INTRODUCTION

The contributions made by knowledge of the human blood groups to fundamental genetics and to the nature and pattern of population variation were enormous. Today over 160 red cell antigens are known, and data on the distribution of their frequencies in various populations of the world were collated by Mourant et al (1954, 1976), and brought up to date by Tills et al (1983). But the greatest advances in population genetics in the last few decades came as a result of the development of methods to identify genes governing other variables such as enzymes and proteins involved in fundamental biological functions in the body. Many enzymes and proteins were found to exist in more than one molecular form. Those multiple forms of proteins arising from genetically determined differences in their primary structure are now termed isozymes or isoenzymes (IUB, 1972). The study of the genetic heterogeneity of isozymes was facilitated by the techniques of electrophoresis. Tiselius (1937) developed this technique as moving boundary electrophoresis, but its first direct application to characterise human gene products was by Pauling and his colleagues (1949) who differentiated by zone electrophoresis the product of the mutant hemoglobin gene for sickle cell (HbS) from normal hemoglobin (HbA). This was a crucial discovery, because it meant that the heterozygote and homozygote could be identified directly by an experimental technique, electrophoresis. In the 1950's the technique of zone electrophoresis was perfected by exploration and development of a variety of supporting media such as agarose, cellulose acetate, starch and polyacrylamide. Its systematic application to the study of human diversity was undertaken by Harris (1969); he examined electrophoretically 18 randomly selected enzyme systems and found that one third of the possible amino acid substitutions can be detected by their difference in electrophoretic migration rate. The data derived from these classical variables became a fundamental tool to test various theoretical models of population genetics and to examine admixture, heterogeneity and migration among different populations of the world (Chakraborty et al 1976; Workman, et al 1973; Wilsenfield and Gajusek, 1976). Further advance has come with the development of new methods of analysis.

IMPROVED METHODS

Isoelectric focusing

The separation of isozymes of the different allelic products by these traditional electrophoretic methods depends on the structure of the protein, the molecular weight, and the interaction of the protein molecule moving in the electric field with a supporting medium at constant pH. However in a given electrophoretic separation, if the external variables such as pH, ionic strength, temperature and field strength are kept constant, it is possible that those isoenzymes which derive from a single amino acid substitution but have similar molecular weight and different net charge, may show similar electrophoretic mobility. Kolin (1955) devised a new approach, using a pH gradient; each protein molecule travels along the gradient till its reaches its iso-electric point. This method of isoelectric focusing (IEF) represented a great advance. The final separation is independent of the external variables, and it allows the differentiation of proteins with a difference of isoelectric point as low as .01 pH unit. The theoretical and methological problems of forming a stable pH gradient were solved by Svensson (1962) and Vesterberg (1969). Using ampholytes to create the pH gradient and polyacrylamide or agarose as a base, an analytical isoelectric focusing method was developed for the comparison of different proteins (Leaback and Rutter, 1968, Fawcett, 1968). This technique has now been used extensively in human population surveys.

Classical variables such as haptoglobin, transferrin, group specific component, alpha 1 antitrypsin, properdin factor B, phosphoglucomutase, and many other systems have now been studied by isoelectric focusing. Both the Tf^{c} and Gc^{1} alleles as detected by starch gel are shown by IEF not to be homogeneous, but instead there are two common suballeles. Both the traditional PGM_1^1 and PGM_1^2 alleles of the enzyme phosphoglucomutase at locus one show two further sub-alleles (PGM_1^{1+}, PGM_1^{1-}, PGM_1^{2+}, PGM_1^{2-}) giving in all 10 common phenotypes. Still greater complexity is found for alpha1 antitrypsin P_i^M, where the common allele can now be differentiated by IEF into five sub-alleles giving fifteen possible phenotypes.

This considerable microheterogeneity demonstrated in various systems has important implications for population genetics. First it suggests that polymorphisms may extend to a greater number of loci, and heterozygosity be more extensive, than was previously postulated. Secondly in taxonomic studies the subtype frequencies provide greater potential for differentiating groups than was possible by the two-allele data from traditional electrophoresis (Papiha et al 1982).

Already many data have been accumulated on these variables from various world populations. The Gc data illustrate the potential usefulness of the new Gc suballeles in differentiating tropical populations.

The data on Gc suballele frequencies from various populations collected by Papiha et al (1985) and other recent studies are listed in Table 1. The African populations generally are distinguished by low Gc*2 gene frequencies (3-12%) whereas populations in the other continents show considerable overlap of frequencies (8-38%); On the classical two-allele electrophoretic model the alternative allele gives similarly restricted information, so that the two-allele data are of limited use in differentiating populations. The subtypes are much more informative, as the following analysis shows. The world populations were divided into nine geographical groups and the Gc suballele frequencies were examined by a discriminant analysis. A computer scatterplot obtained using Gc*1F and Gc*2 alleles is presented in Figure 1 and the predicted group membership in Table 2.

The lowest frequency of the Gc*1F gene characterises the Basques, thought to be the indigenous population of Europe, and they are well outside the European cluster. The populations from Africa and Southeast Asia have a very high frequency of the Gc*1F allele; although their clusters overlap the cross group membership for these populations is only some 23%. East Asian populations show a slightly lower range of Gc*1F allele frequencies (42-67%) than the African (55-86%), but the Gc*1F gene frequency differentiates both of these from the Caucasian groups (European, South west, South and Central Asia), amongst all three of which there is considerable overlap of the Gc*1F frequency ranging from 8-35%. The predicted group membership of the South and North American Indians also shows considerable spread but the frequencies in the majority of populations resemble those in European and Eskimo populations (50 and 25% respectively).

In two-allele systems, genetic differentiation of subgroups, as measured

Table 1: Gc subtype gene frequencies

Population	Gc1*F	Gc*1S	Gc*2	Number tested	Reference
South, Central and North American Indians					
Machiguenaga, Peru	0.106	0.603	0.289	180	Matsumoto et al, 1980
Quechuan, Peru	0.278	0.598	0.119	97	Matsumoto et al, 1980
La Minita, Mexico	0.355	0.504	0.141	128	Dykes et al, 1983
Chamizal, Mexico	0.348	0.480	0.172	102	Dykes et al, 1983
Apache-San Carlos	0.318	0.596	0.086	457	Dykes et al, 1983
Apache - White River	0.295	0.669	0.036	127	Dykes et al, 1983
Blackfeet	0.189	0.615	0.196	74	Dykes et al, 1983
Cocopa	0.370	0.445	0.185	135	Dykes et al, 1983
Maricopa	0.390	0.426	0.184	68	Dykes et al, 1983
Navaho	0.403	0.529	0.068	103	Dykes et al, 1983
Pima	0.449	0.410	0.141	332	Dykes et al, 1983
Walapai	0.348	0.574	0.078	115	Dykes et al, 1983
Yakima	0.310	0.560	0.130	92	Dykes et al, 1983
Dogrib, Canada	0.380	0.551	0.070	107	Szathmary et al, 1983
Minneapolis, U.S.A.	0.153	0.573	0.274	3482	Dykes et al, 1983
Mennonites, U.S.A.	0.118	0.567	0.315	611	Dykes et al, 1983
Minnesota, U.S.A.	0.155	0.566	0.279	7247	Dykes et al, 1983
Eskimo, Alaska	0.267	0.492	0.189	328	Matsumoto et al, 1980
Eskimo, Alaska	0.261	0.345	0.334	307	Dykes et al, 1983
Tuvintisi, Siberia	0.522	0.284	0.194	227	Dykes et al, 1983
Europe					
West of Scotland	0.153	0.555	0.292	1675	Baxter & White, 1984
Hessen, Germany	0.141	0.598	0.261	680	Kuhn et al, 1978
Berlin, Germany	0.143	0.584	0.273	251	Martin, 1979
Munich, Germany	0.140	0.598	0.261	1523	Weidinger et al, 1984
Dusseldorf, Germany	0.156	0.548	0.296	1157	Schell et al, 1980
Tuscany, Italy	0.148	0.594	0.256	965	Bargagna et al, 1983
Padua, Italy	0.159	0.561	0.277	732	Corturo et al, 1983
Viterbo, Italy	0.153	0.561	0.286	245	Petrucci et al, 1983
Rome, Italy	0.158	0.591	0.251	397	Petrucci et al, 1983
Frosmone, Italy	0.145	0.541	0.314	298	Petrucci et al, 1983
Latina, Italy	0.113	0.601	0.286	301	Petrucci et al, 1983
Rhone	0.146	0.562	0.291	989	Scheffrahn, 1983
Sweden	0.139	0.606	0.253	339	Svensson & Hjalmansson, 1981
Iceland	0.107	0.631	0.262	385	Karlson et al, 1983a

Population	Gc1*F	Gc*1S	Gc*2	Number tested	Reference
Middle & East Africa					
Arab, Israel	0.212	0.602	0.186	342	Nevo & Cleve, 1983
Tunisia (mixed origin)	0.260	0.525	0.215	349	Lefranc et al, 1981
North & Central American Blacks					
Black Caribs:					
St. Vincent, W. Indies,	0.498	0.360	0.126	311	Dykes et al, 1983
Guatemala, Livingston	0.637	0.256	0.107	215	Dykes et al, 1983
Stann Creek	0.536	0.312	0.152	274	Dykes et al, 1983
Belize	0.553	0.304	0.143	217	Dykes et al, 1983
USA Blacks, Minnesota	0.678	0.186	0.106	540	Dykes et al, 1983
Asia					
Lucknow	0.111	0.627	0.262	63	Papiha et al, 1983
Orissa	0.300	0.544	0.144	80	Papiha, 1983
Konda Kamma, Rajavommangi	0.271	0.525	0.204	126	Walter et al, 1981
Konda Kamma, Maredumilli	0.222	0.444	0.333	108	Walter et al, 1981
Koya Dora, Maredumilli	0.201	0.582	0.217	97	Walter et al, 1981
Koya Dora, W. Godavari	0.133	0.678	0.189	177	Walter et al, 1981
Koya Dora, Warangal	0.221	0.636	0.143	70	Walter et al, 1981
Madiga, Warangal	0.233	0.598	0.169	133	Walter et al, 1981
Lainbadi, Khamman	0.146	0.573	0.231	48	Walter et al, 1981
Punjabis, Birmingham	0.191	0.519	0.290	243	Karlson et al, 1983
Indians, N. Sumatra	0.141	0.603	0.255	78	Tan et al, 1981
Fukushima, Japan	0.485	0.240	0.247	893	Abe, 1983
Shimane, Japan	0.452	0.245	0.275	600	Yuosa et al, 1984
Yamaguchi, Japan	0.442	0.245	0.280	400	Yuosa et al, 1984
Osaka, Japan	0.456	0.258	0.254	342	Shibata et al, 1983
Japan	0.421	0.301	0.258		Matsumoto et al, 1979
Okinawa, Japan	0.473	0.209	0.316	502	Matsumoto et al, 1980
Hokkaido, Ainu	0.579	0.203	0.209	271	Matsumoto et al, 1980
Chinese, Taiwan	0.397	0.271	0.303	373	Matsumoto et al, 1980
Atyal, Taiwan	0.588	0.294	0.110	354	Matsumoto et al, 1980
Conjj, Korea	0.434	0.234	0.304	303	Matsumoto et al, 1980
Indonesian, Java	0.534	0.281	0.176	176	Matsumoto et al, 1980
Kadazan, Borneo	0.610	0.264	0.123	260	Matsumoto et al, 1980

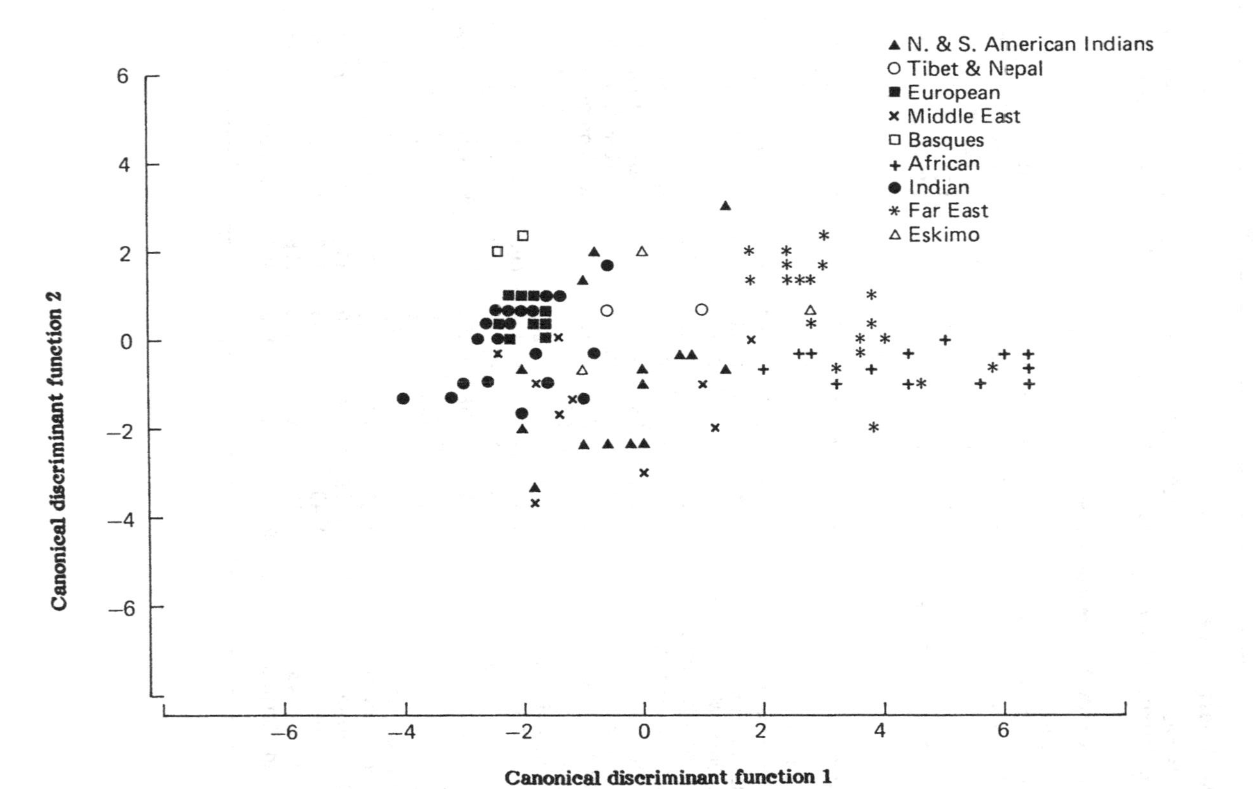

Figure 1: Discrimination of populations by Gc*1F and Gc*2 allele frequencies

Table 2: Predicted group membership

Group	Group No.	No. of Popula-tions		1	2	3	4	5	6	7	8	9
North & South American Indians	1	18	No.:	2	2	1	8	0	0	1	0	4
			%:	11.1	1.1	5.6	44.4	0.0	0.0	5.6	0.0	22.2
Tibet & Nepal	2	2	No.:	0	1	0	0	0	0	0	0	1
			%:	0.0	50.0	0.0	0.0	0.0	0.0	0.0	0.0	50.0
European	3	32	No.:	0	0	25	0	0	0	7	0	0
			%:	0.0	0.0	78.1	0.0	0.0	0.0	21.9	0.0	0.0
South-West Asia	4	12	No.:	2	0	0	5	0	0	44	0	1
			%:	16.7	0.0	0.0	41.7	0.0	0.0	33.3	0.0	8.3
Basques	5	2	No.:	0	0	0	0	2	0	0	0	0
			%:	0.0	0.0	0.0	0.0	100.0	0.0	0.0	0.0	0.0
African	6	13	No.:	0	0	0	0	0	10	0	3	0
			%:	0.0	0.0	0.0	0.0	0.0	76.9	0.0	23.1	0.0
Indian	7	23	No.:	1	1	10	4	0	0	7	0	0
			%:	4.3	4.3	43.5	17.4	0.0	0.0	30.4	0.0	0.0
Far East	8	24	No.:	0	0	0	0	0	6	0	16	2
			%:	0.0	0.0	0.0	0.0	0.0	25.0	0.0	66.7	8.3
Eskimo	9	3	No.:	1	1	0	0	0	0	0	1	0
			%:	33.3	33.3	0.0	0.0	0.0	0.0	0.0	33.3	0.0

by Wright's F_{ST} values, will be the same for each allele. Incorporation of the subtypes and the third allele obviously enhances the detectable heterogeneity among populations. A recent study of genetic differentiation showed a wide range of F_{ST} values in different regions of the world, the lowest being in Asiatic Indians (2%) and highest for African populations (26%). Overall the genetic differentiation (F_{ST}) for the Gc locus as studied by IEF was ten times that previously reported using conventional electrophoresis (Tills, 1977; Papiha et al, 1985).

The exact biological function of the Gc protein is still not known, but since Gc of human plasma binds vitamin D it is postulated that this acts as the carrier protein of vitamin D. Since the synthesis of vitamin D is promoted by sunlight, Mourant et al (1976) studied the distribution of Gc*2 allele frequencies in relation to insolation. Overall the world frequencies of Gc*2 were found to be inversely related to the degree of insolation but there were many outliers. In the present analysis, average frequencies of Gc*1F from different regions of the world were plotted together with data on solar radiation. It is quite evident that tropical regions of Africa and south-east Asia with high insolation (average global solar radiation 25.3 $mWem^{-2}$/24hr and 20 $mWem^{-2}$/24hr respectively) have very high Gc*1F frequencies. In general, populations near to the equator tend to show a higher frequency of Gc*1F and the frequency tends to decrease as one moves towards high latitude. To this overall geographical association the chief exceptions other than recent migrants, seem to be the Siberian populations which show a higher 1F* gene frequency than expected. The number of populations as yet studied is small, but this preliminary analysis by mapping and genome differentiation procedures suggests that for the Gc locus there may be some selective advantage in tropical and semitropical climates, particularly in relation to the Gc*1F suballele.

In addition to the complexity demonstrated for these common alleles, this powerful separation procedure has revealed many rare variants. Some of these new mutants have restricted distributions and may therefore be useful for taxonomic differentiation of human populations. For the Gc system traditional gel electrophoresis on starch or polyacrylamide was used successfully to identify rare genetic variants such as Gc X, Gc Y, Gc Ab, Gc Chip, Gc Z, Gc Bangkok, Gc Norway, Gc Eskimo. Two-dimensional electrophoresis (polyacrylamide and crossed immunoelectrophoresis) and immunofixation procedures identified further variants, Gc D, Gc W, Gc J, Gc op, Gc B, Gc T and Gc Toulose. The characterisation of Gc variants by IEF started in 1977, and by 1978 when the International Workshop on the Gc system was held 30 rare Gc mutant alleles were already identified (Constans & Cleve, 1979).

The nomenclature of these mutant variants was standardised according to their electrophoretic mobility. The anodal variants were termed Gc1A or Gc2A and the cathodal variants Gc1C or Gc2C. Variants described by traditional electrophoresis are now designated by number e.g. GcAb (Gc1A1) and GcChip (Gc1A10). This facilitates the comparison of the new mutants in various populations and their applications in anthropological and related studies.

Recent isoelectric focusing on polyacrylamide gels containing 3M Urea has further improved separation and disclosed differences which are not apparent by simple IEF. Urea changes the secondary and tertiary structures of protein molecules, by disrupting the weak ionic links and hydrogen bonds. The molecules thus unfolded may reveal charge differences between proteins not expressed in the native state. The diagrammatic separation of the cathodal variants is shown in Figure 2. The cathodal variants Gc 1C13 and 1C34 show similar mobility in ordinary IEF gels, but when compared in 3M urea the 1C34 mutant band is found cathodal to the 1C13 mutant band (Fig. 3). A total of 84 rare variants has been identified in the Gc system, in addition to the three common alleles (1F, 1S, 2). These include twenty two Gc1A, thirty seven Gc1C, fourteen Gc2A and eleven Gc2C alleles (Constans et al 1983).

High resolution analytic IEF with immobiline

The method of isoelectric focusing which utilises carrier-ampholytes for the generation of pH gradients has been widely used in population genetics. This ampholytic base technique has certain inherent limitations, the most significant being cathodic drift and interaction of carrier ampholytes with peptides of similar nature (Fawcett, 1975; Righetti & Chillemi, 1978).

A system consisting of a series of buffering acrylamide derivatives was introduced which co-polymerise with the acrylamide and bisacrylamide in order to generalise the immobilized pH gradient. Immobilines of two different pH values, between which the separation of the protein occurs, are selected and these are mixed separately with polyacrylamide and other chemicals to obtain two solutions differing in pH. These solutions are then poured with the help of a gradient mixer to obtain a polyacrylamide gel containing an immobilised pH gradient.

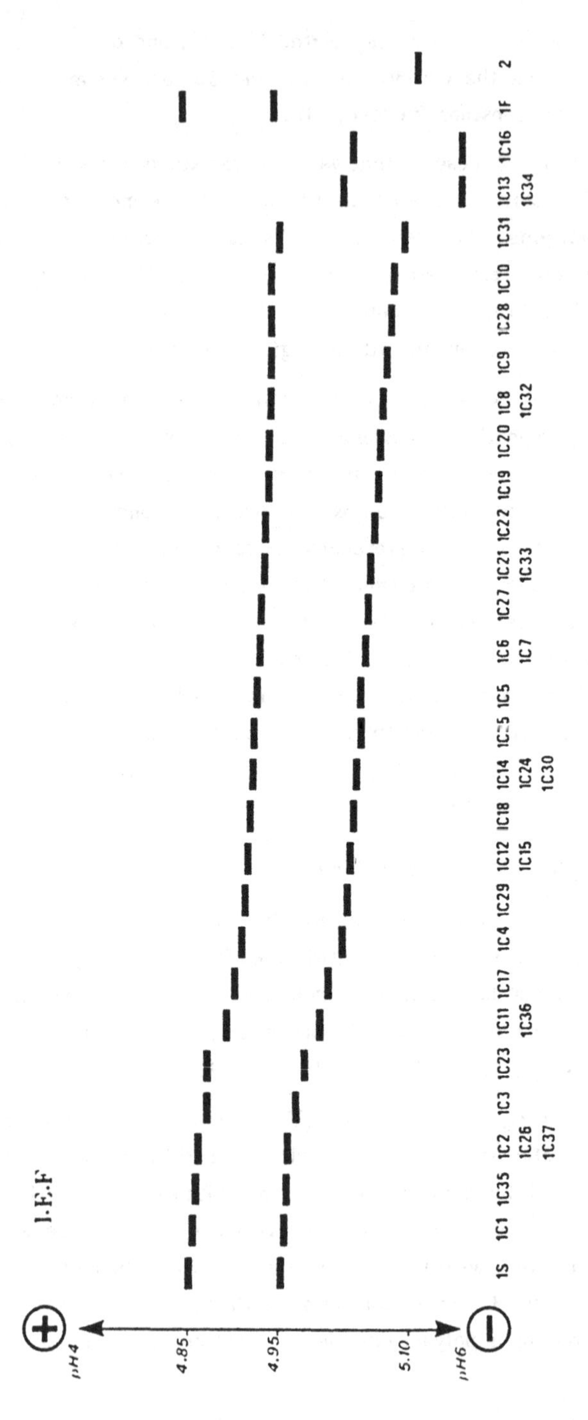

Figure 2: Diagrammatic representation of cathodal variants of Gc

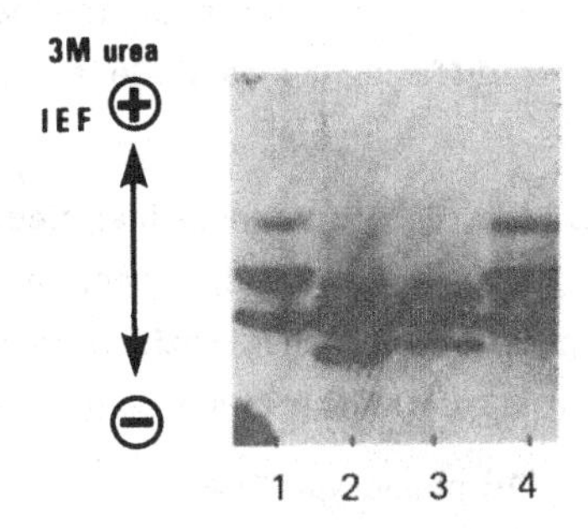

Figure 3: Refined separation of IC34 and IC13 in 3M urea (samples 2 and 3, respectively)

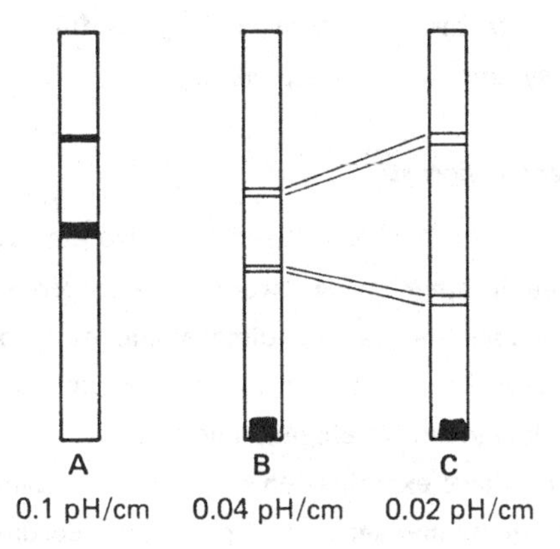

Figure 4: The PiM1M3 subtype by conventional IEF (A) and using two different pH slopes with immobiline systems (B and C) (After Gorg et al, 1983.)

There are a number of technical advantages to the use of an immobiline system but a fundamental advantage for population study is that the narrow range gradient can be expanded into a wider space, and thus provide a wider corridor between the two electrophoretic bands of the heterozygote. All proteins with isoelectric points differing by as little as .001 pH unit can be successfully resolved.

The application of this high resolution analytical electrofocusing with an immobiline pH gradient has proved very useful, especially for the system of alpha1-antitrypsin. The separation of alpha1 antitrypsin in ordinary IEF and with an immobiline pH gradient is shown diagrammatically in Figure 4.

The alpha 1 antitrypsin (Pi) phenotype PiM1M3 was analysed in the same thick polyacrylamide gel but with pH gradients of different slopes. By conventional IEF, in the subtype PiM1M3 the bands representing M1 and M3 cannot be clearly distinguished from each other. By using an immobilized pH gradient with a slope of 0.04 pH unit/cm, the two bands are clearly separated. Change of the slope to .02 pH unit/cm, by selecting immobilines of two different pH ranges, increases the corridor between the M1 and M3 bands, By the use of this immobiline pH gradient new mutant alleles (M3, M, M5) have been clearly distinguished. Such separation is difficult by ordinary IEF. The application of immobiline gel gradients has also proved very useful in analysing the polymorphisms of the Tf system (Gorg et al, 1983).

Two-Dimensional Electrophoresis

In 1956, Smithies and Poulik realised that two electrophoretic separations at right angles to each other using two different properties of the protein produce better resolutions than a one dimensional electrophoresis. O'Farrell (1975), using IEF in one dimension and sodium dodecyl sulphate (SDS) gradient PAGE in the other, developed an elegant and powerful technique of 2-D PAGE electrophoresis which allows examination of several thousand gene products in a single gel. This technique separates protein according to differences in isoelectric point and molecular weight. With the conditions employed, more than two hundred serum peptides can be resolved by 2D electrophoresis.

Several allele products have been studied by 2D gels in population surveys of man. The average heterozygosity using 2D electrophoresis was estimated as

1-3 percent (McConkey et al, 1979; Rosenblum et al, 1983), which is significantly lower than the 6.3% estimated by Harris et al (1977) using traditional 1D electrophoresis. While these differences in estimates may reflect the different classes of proteins examined in the two studies, for the same enzymes and proteins studied by the two methods it is possible that the experimental conditions of 2D electrophoresis may not be ideal for revealing the same genetic heterogeneity. Wanner et al (1982) showed that for the study of human mutation rate the sensitivity of the 2D system may be slightly lower than 1D, but if use is restricted to the well defined central region of the gel, the resolution offered by 2D electrophoresis has the sensitivity required for the study of mutations.

Two additional advantages of the 2-D system make the technique a particularly valuable development. First only soluble proteins can be analysed by 1-D electrophoresis, whereas membrane proteins and organelle components also can be examined by 2-D electrophoresis. Recent analysis of the protein of the human cells and tissue with 2-D electrophoresis revealed three polymorphisms at autosomal loci responsible for the synthesis of polypeptide 31K, LC64K and C100K (Kondo et al, 1984; Hamaguchi et al, 1982a,b). These polymorphisms could not be demonstrated by traditional electrophoresis or IEF. The understanding of genetic variation of these abundant cellular proteins is essential because these peptides play important roles in the structure and the complex regulatory functions of the cell.

Secondly 2-D electrophoresis allows a broad-based search for possible disease-related protein markers which may possibly be the basis for new diagnostic tests. This system has been used to subtype myeloma proteins, and to determine the isoelectric points of the Ig-light chains which may be of value in predicting a patient's risk of developing renal damage (Beaufils & Morel-Maroger, 1978). The systematic use of 2D in developing clinical laboratory tests has recently been discussed (Young and Tracy, 1983).

CONCLUSION

In conclusion it appears that the relatively new electrophoresis technique of isoelectric focusing is an excellent tool for the study of genetic diversity in populations of various species. The introduction of immobiline pH gradients and detergents has further expanded its potential and usefulness. By contrast 2D electrophoresis due to its laborious, time consuming and non-specific nature does not appear as yet to be of direct application in routine analysis of populations, although a large number of cytosol and tissue proteins have been examined by it and many have been found to be polymorphic. The biological functions of these 2-D defined polymorphisms are still to de elucidated. But these additional polymorphic proteins will be of particular value as genetic markers in pedigree analysis and gene mapping, and it will also be essential to include them in attempts to estimate the true heterozygosity of the human genome.

REFERENCES

Abe, S. (1983). Genetic polymorphism of red cell enzymes (AcP, EsD, GPT, G-PGD) and of serum protein (Gc) in Fukushira prefecture. Japanese Journal of Human Genetics, **28**, 196-200.

Baxter, M. & White, I. (1984). A method for the identification and typing of the subtypes of the Gc1 allele from dried blood stains. Journal of the Forensic Science Society, p. 24.

Beaufils, M. & Morel-Maroger, L. (1978). Pathogens of renal disease in monoclonal gammopathies: current concepts. Nephrone, **20**, 125-131.

Bargagna, M., Domenici, R. & Giari, A. (1983). Distribution of Gc, P1 and Tf subtypes by isoelectric focusing in Tuscany. Proceedings of the 10th International Congress of the Society for Forensic Haemogenetics, Munchen, pp. 391-386.

Chakraborty, R., Blanco, R., Rothhammer, F. & Llop, E. (1976). Genetic variability in Chilean Indian populations and its association with geography, language and culture. Social Biology, **23**, 73-81.

Constans, J. & Cleve, H. (1979). Group-specific component. Report of the First International Workshop. Human Genetics, **48**, 143-149.

Constans, J., Cleve, H., Dykes, D., Fischer, M., Kirk, R.L., Papiha, S.S., Scheffran, W., Scherz, R., Thymann, M. & Weber, W. (1983). The polymorphism of vitamin D-binding protein (Gc). Isoelectric focusing in 3M urea as an additional method for identification of genetic markers. Human Genetics, **65**, 176-180.

Cortivo, P., Biasiolo, M., Scrorretti, C. & Bencioline, P. (1983). Transferrin, alpha-1-antitrypsin, group-specific component and phosphoglucomutase 1 frequency and distribution in the population of Padua by isoelectric focusing. Proceedings of the 10th International Congress of the Society for Forensic Haemogenetics, Munchen, pp. 419-437.

Dykes, D., Polesky, H. & Cox, E. (1981). Isoelectric focusing of Gc (vitamin D binding globulin) in parentage testing. Human Genetics, **58**, 174-175.

Dykes, D.D., Crawford, M.J. & Polesky, H.F. (1983). Population distribution in north and central America of PGM_1 and Gc subtypes as determined by isoelectric focusing (IEF). American Journal of Physical Anthropology, **62**, 137-145.

Fawcett, J.S. (1968). Isoelectric fractionation of proteins on polyacrylamide gels. FEBS letter, **1**, 81.

Fawcett, J.S. (1975). In: P.G. Righetti (ed.), Isoelectric Focusing and Isotachophoresis, pp. 25-37. Amsterdam: North Holland/American Elsevier.

Gorg, A., Postel, W., Weser, J., Weidinger, S., Patutschnick, W. & Cleve, H. (1983). Isoelectric focusing in immobilized pH gradients for the determination of the genetics of Pi(α_1-antitrypsin) variants. Electrophoresis, **4**, 153-157.

Hamaguchi, H., Yamada, M., Noguchi, A., Fujii, K., Shibasaki, M., Mukai, R., Yabe, T. & Kondo, I. (1982a). Genetic analysis of human lymphocyte protein by two-dimensional gel electrophoresis: II. Genetic polymorphism of lymphocyte cytosol 64k polypeptide. Human Genetics, **60**, 176-180.

Hamaguchi, H., Yamad, M., Shibasaki, M. & Kondo, I. (1982b). Genetic analysis of human lymphocyte proteins by two-dimensional gel electrophoresis. IV. Genetic polymorphism of cytosol 100k polypeptide. Human Genetics, **62**, 148-151.

Harris, H. (1969). Enzyme and protein polymorphism in human populations. British Medical Bulletin, **25**, 5.

Harris, H., Hopkinson, D.A. & Edward, Y.H. (1977). Polymorphism and the subunit structure of enzymes: a contribution to the neutralist - selectionist controversy. Proceedings of the National Academy of Sciences, U.S.A. **74**, 698-701.

Karlsson, S., Skaftadottir, I., Aonason, A., Thordarson, G., Jenson, O. (1983a). Gc subtypes in Icelanders. Human Heredity, **33**, 5-8.

Karlsson, S., Skaftadottir, I., Aonason, A., Makintosh, P. & Jenson, O. (1983b). Gc subtypes in Northern India. Human Heredity, **33**, 199-200.

Kolin, A. (1955). Isoelectric spectra and mobility spectra: a new approach of electrophoretic separation. Proceedings of the National Academy of Sciences, U.S.A. **41**, 101-110.

Kondo, I., Yamakawa, K., Shibasaki, M., Yamamoto, T. & Hamaguchi, H. (1984). Genetic analysis of human lymphocyte proteins by two-dimensional gel electrophoresis and genetic polymorphism of cytosol 31k polypeptide. Human Genetics **66**, 244-247.

Leabach, D.H. & Rutter, A.C. (1968). Polyacrylamide isoelectric focusing. A new technique for the electrophoresis of proteins. Biochemical and Biophysical Research Communications, **32**, 447-453.

Lefranc, M.P., Chibani, J., Helal, A.N., Boukef, K., Seger, J. & Lefranc, G. (1981). Human transferrin (Tf) and group-specific component (Gc) subtypes in Tunisia. Human Genetics, **59**, 60-63.

McConkey, E.H., Taylor, B.J. & Phan, D. (1979). Human heterozygosity: a new estimate. Proceedings of the National Academy of Sciences, U.S.A., **76**, 6500-6504.

Martin, W. (1979). Zur Gc und Tf Typisierung mit Hilfe der Isoelektrischen Fokussierung. 8 International Tagung der Gesellschaft fur Forensische Blutgruppenkude, London, p. 507.

Matsumoto, H., Matsui, K., Ishida, N., Ohkura, K. & Teng, Y.S. (1980). The distribution of Gc subtypes among the mongoloid populations. American Journal of Physical Anthropology, **53**, 505-508.

Matsumoto, H., Toyomasu, T., Tamaki, Y., Katayma, K. & Matsiu, K. (1979). The distribution of Gc subtypes in Japanese and its application in paternal cases. Japanese Journal of Legal Medicine, **33**, 74-79.

Mourant, A.E. (1954). The Distribution of the Human Blood Groups. Oxford: Blackwell.

Mourant, A.E., Kopec, A.C. & Domaniewska-Sobewzka, K. (1976). The Distribution of the Human Blood Groups and Other Polymorphisms. London: Oxford University Press.

Nevo, S. & Cleve, H. (1983). Gc subtypes in the middle east: report on an Arab Moslem population from Israel. American Journal of Physical Anthropology, **60**, 49-52.

O'Farrell, P.H. (1975). High resolution two-dimensional electrophoresis of proteins. Journal of Biological Chemistry, **250**, 4007-4021.

Papiha, S.S. (1983). Phosphoglucomutase (PGM_1) and group-specific component (Gc) subtypes in the Langia Soura tribe of Orissa, India. Journal of the Indian Anthropological Society, **18**, 39-43.

Papiha, S.S., Roberts, D.F., White, I., Chahal, S.M.S. & Asefi, J.A. (1982). Population genetics of the group-specific component (Gc) and phosphoglucomutase (PGM_1) studied by isoelectric focusing. American Journal of Physical Anthropology, **59**, 1-8.

Papiha, S.S., Agarwal, S.S. & White, I. (1983). Association between phosphoglucomutase (PGM) and group-specific component (Gc) subtypes and tuberculosis. Journal of Medical Genetics, **20**, 220-222.

Papiha, S.S., Constans, J., White, I. & McGregor, I.A. (1985). Group specific component (Gc) subtypes in Gambian and Transkeian populations: a description of a new variant. Annals of Human Biology, **12**, 17-26.

Pauling, L., Itano, H.A., Singer, S.J. & Wells, I.C. (1949). Sickle cell anaemia, a molecular disease. Science, **110**, 543-544.

Petrucci, R. & Congedo, P. (1983). Genetic studies of Gc (vitamin D binding globulin) polymorphism is the population of Latium (Italy). Journal of Human Evolution, **12**, 439-441.

Righetti, P.G. & Chillemi, F.J. (1978). Isoelectric focusing of peptides. Journal of Chromatography, **157**, 243-251.

Rosenblum, B.B., Neel, J.V. & Hariash, S.M. (1983). Two-dimensional electrophoresis of plasma polypeptides reveals high heterozygosity indices. Proceedings of the National Academy of Sciences, U.S.A., **80**, 5002-5006.

Scheffrahn, W. (1983). Gc variants in Swiss populations. Proceedings of the 10th International Congress of the Society for Forensic Haemogenetics, Munchen, pp. 415-417.

Scheil, H.G., Driesel, A.J. & Rohrborn, G. (1980). Distribution of Gc subtypes in Western Germany (Dusseldorf region). Zeitschrift Richtsmedizin, **84**, 95-97.

Shibata, K. (1983). Haptoglobin, group-specific component, transferrin and alpha 1 antitrypsin subtypes and new variants in Japanese. Japanese Journal of Human Genetics, **28**, 17-27.

Svensson, H. (1962). Isoelectric fractionation, analysis and characterization of ampholytes in natural pH gradient III. Description of apparatus for electrolysis in columns stabilized by density gradient and direct gradients and direct determination of isoelectric point. Archives of Biochemistry and Biophysics, Supplement 1, 131-142.

Svensson, M. & Hjalmarsson, K. (1981). Distribution of Gc subtypes by isoelectric focusing in Sweden. 9 International Tagung der Gescellschaft fur Forensische Blutgruppenkunde, Bern, p. 559.

Tan, S.G., Gan, Y.U., Asuan, K. & Abdullah, F. (1981). Gc subtyping in Malaysians and in Indonesians from North Sumatra. Human Genetics, **59**, 75-76.

Tills, D. (1977). The use of F_{ST} statistic of Wright for estimating the effects of genetic drift, selection and migration in populations with special reference to Ireland. Human Heredity, **27**, 153-159.

Tills, D., Kopec, A.C. & Tills, R.E. (1982). The Distribution of the Human Blood Groups and Other Polymorphisms. Supplement 1, Oxford Monographs on Medical Genetics. Oxford University Press.

Tiselius, A. (1937). A new apparatus for electrophoresis. Analysis of colloidal mixtures. Transactions of the Faraday Society, **33**, 524-528.

Vesterberg, O. (1969). Synthesis and isoelectric fraction of carrier ampholytes. Acta Chemica Scandinavica, **23**, 2653-2663.

Walter, H., Pahl, K-P., Hilline, M., Veeraju, P., Goud, J.D., Naidu, J.M., Babu, M., Jai, G. & Kisham, G. (1981). Genetic markers in eight endogamous population groups from Andhra Pradesh (South India). Zeitschrift fur Morphologie und Anthropologie, **72**, 325-338.

Wanner, L.A., Neel, J.V. & Meisler, M.H. (1982). Separation of allelic variants by two-dimensional electrophoresis. American Journal of Human Genetics, **34**, 209-215.

Weidinger, S., Cleve, H., Schwarzfischer, F. & Patutsch, R. (1981). The Gc system in paternity examinations: application of Gc subtyping by isoelectric focusing. In: K. Hummel & J. Gercow (eds.), Biomathematical Evidence of Paternity, pp. 113-121. Berlin: Springer Verlag.

Wiesenfeld, S.L. & Gadjusek, D.C. (1976). Genetic structure and heterozygosity in the Kuru region, Eastern Highlands of New Guinea. American Journal of Physical Anthropology, **45**, 177-190.

Workman, P.L., Harpending, H.C., Lalouel, J.M., Lynch, C., Niswander, J.D. & Singleton, R. (1973). Population studies on south-western Indian tribes, VI: Papago population structure, a comparison of genetic and migration analyses, In: N.E. Morton (ed.), Genetic Structure of Populations, pp. 166-194. University of Hawaii Press.

Young, D.S. & Tracy, R.P. (1983). Two-dimensional electrophoresis in the development of clinical laboratory tests. Electrophoresis, **4**, 117-121.

Yuosa, I., Saneshige, Y., Okamoto, N., Ikawa, S., Ikebuchi, J., Hikita, T., Inoue, T. & Okado, K. (1983). Distribution of Hp, Tf, Gc and Pi polymorphisms in a Nepalese population. Human Heredity, **33**, 302-306.

Yuosa, I., Suenga, K., Gotoh, Y., Ito, K. & Yokoyama, N. (1984). Gc types in western Japan: report of a new variant Gc IC35. Human Heredity, **34**, 174-177.

HLA VARIATION IN THE TROPICS

J. WENTZEL

Department of Human Genetics, University of Newcastle upon Tyne, Newcastle upon Tyne, U.K.

INTRODUCTION

The HLA system consists of over 100 antigens in six segregant series. Kissmeyer-Nielsen (1968) first identified two closely linked loci, with genes controlling two independent series of antigens, the A and B series, while Sandberg et al (1970) proposed a third locus (C). These are situated close together on chromosome 6 (Fig. 1), but their independence was confirmed by the cross-overs that occur between them (Kissmeyer-Nielsen et al, 1969; Low et al, 1974). Whereas antigens at these loci are recognised by serological methods, the fourth locus (D) was recognised from examination of mixed lymphocyte reactions. When lymphocytes from two individuals are mixed together in appropriate culture conditions, the one stimulates the other to blastoid transformation and proliferation. The extent of the reaction depends upon the degree of histocompatibility between the individuals, and is largely dependent on the D locus. Serological correlates of the MLR were therefore sought, and following the detection on leukaemic cells of B cell alloantigens (Walford et al, 1975), a series of DR (= D-related) antisera were standardised. Extremely close to the DR locus is one controlling a second series (DQ) of antigens (Duquesnoy et al, 1979; De Kretser et al, 1983), while a further locus (DP) governing reactivity in the primed lymphocyte test (secondary MLC) is shown by both cellular and serological techniques (Shaw et al, 1980, 1982; Mawas et al, 1980). The relationship between the specificities defined by MLC and the DR, DP and DQ antigens remains unresolved.

Antigens of the A, B and C series are known as class 1 antigens, and are found on the cells of most body tissues. Increased knowledge of their molecular structure has led to more precise knowledge of their genetic control. Each class 1 antigen is composed of two polypeptide chains; the α chain is a glycoprotein of MW 43000, its allospecificity resides in its distal portion, and the genes map to chromosome 6. The β chain is composed of β-2 microglobulin, and it maps to chromosome 15. The other antigens are known as class 2 and are only found on B lymphocytes, some T cells, and specialised antigen-presenting cells. Of the class 2 antigens, DR molecules are composed of two noncovalently linked glycoproteins, with α chains of MW 33000 and β chains 29000; DP and DQ are also two-chain structures of similar MW to DR. In all class 2 antigens, both the α and β chains map to chromosome 6 where theD region contains six alpha-chain genes (Auffray et al, 1984; Spelmann et al, 1984), and at least seven beta-chain genes (Kratzin et al, 1981; Mach et al, 1984).

To date the numbers of alleles known respectively in the A, B, C, DR, DP and DQ are 19, 37, 8, 12, 3 and 6. The system is therefore highly polymorphic at each locus (Table 1).

There are very few population studies published on the DQ or DP antigens so far, but sufficient has been done on A, B, C and DR for some distinct patterns to have emerged (Table 2). Thus, A1, A25, B37, B38 and Bw63 are almost exclusively confined to Europeans; Aw34, Aw36, Aw43, Bw42, Bw53 and Bw58 to Negroes; and Bw52, Bw54 and Bw61 to Orientals. Amerindians are similar to Orientals but apparently lack Bw54 (which is a marker for Orientals) as well as A26, A11, B7, B13 and B44. A28 and B39 are present in Amerindians but absent from Orientals. There appear to be no HLA antigens that are specific to Amerindians (Pickbourne et al, 1977; Baur & Danilovs, 1980).

MATERIAL AND METHODS

Data in the literature on a number of populations were examined for diversity of HLA antigen frequencies. From these frequency matrices the genetic distances and kinship between the populations were examined by the method of Harpending and Jenkins (1973). The principal components of the kinship matrix (R) were extracted, and relationships of the populations shown by plotting their positions on the first two eigenvectors. The alleles mainly contributing to these positions were identified by reference to the eigenvectors of the scaled matrix of covariances among allele frequencies (S).

RESULTS

(a) *Intercontinental*

In the first analysis a selection of populations was chosen to represent different continental groups; there were two groups of Europeans, two of

Table 1: HLA antigen specificities as defined by the 9th International Histocompatibility Workshop

HLA-A		HLA-B			HLA-C	HLA-D		HLA-DR		HLA-DQ	HLA-DP
A1		B5	B51	Bw4	Cw1	Dw1		DR1	DRw52	DQw1	DPw1
A2			Bw52	Bw6	Cw2	Dw2		DR2	DRw53	DQw2	DPw2
A3		B7			Cw3	Dw3		DR3		DQw3	DPw3
A9	A23	B8			Cw4	Dw4		DR4			DPw4
	A24	B12	B44		Cw5	Dw5		DRw5	DRw11		DPw5
A10	A25		B45		Cw6	Dw6	Dw18		DRw12		DPw6
	A26	B13			Cw7		Dw19	Dw6	DRw13		
	Aw34	B14	Bw64		Cw8	Dw7	Dw11		DRw14		
	Aw66		Bw65				Dw17	DR7			
A11		B15	Bw62			Dw8		DRw8			
Aw19	A29		Bw63			Dw9		DRw9			
	A30	B16	B38			Dw10		DRw10			
	A31		B39			Dw12					
	A32	B17	B57			Dw13					
	Aw33		B58			Dw14					
A28	Aw68	B18				Dw15					
	Aw69	B21	B49			Dw16					
Aw36			Bw50								
Aw43		Bw22	Bw54								
			Bw55								
			Bw56								
		B27									
		B35									
		B37									
		B40	Bw60								
			Bw61								
		Bw41									
		Bw42									
		Bw46									
		Bw47									
		Bw48									
		Bw53									
		Bw59									
		Bw67									
		Bw70	Bw71								
			Bw72								
		Bw73									

Specificities to the left of brackets include specificities to the right of those brackets.
Bw4, Bw6, DRw52 and DRw53 are 'supertypic' specificities.
Dw1, 2, 3, 4, 7 and 8 correlate strongly with the DR specificities having the same numbers. Dw5 correlates with DRw11, Dw6 with DRw13, Dw9 with DRw14.

Table 2: Racial distribution of HLA antigen specificities

	European	Negroid	Oriental
Common	A1, A25, B37 Bw38, Bw63	Aw34, Aw36, Aw43 Bw42, Bw53, Bw58	Bw52, Bw54, Bw61
Rare	Aw34, Aw36, Aw43 Bw42, Bw52, Bw53 Bw54, Bw58, Bw59 Bw61	A1, A11, A25 B13, B37, Bw38 Bw51, Bw52, Bw54 Bw55, Bw56, Bw60 Bw61, Bw62, Bw63 Cw1	A1, A3, Aw23 A25, A28, A29 Aw30, Aw32, Aw34 Aw36, Aw43, B8 B14, B18, B27 B37, Bw38, Bw41 Bw42, Bw45, Bw49 Bw50, Bw53, Bw57 Bw58, Bw63, Cw2 Cw5, Cw6, Cw8 DR3, DR7

Japanese, two of American Indians, two of African Blacks, and one of American Blacks. One of each pair and the American Blacks were the pools regarded as representative by the 7th International Histocompatibility Workshop, and the remainder were drawn from the literature. For these groups, frequencies were available of 14 alleles at the HLA-A locus and 19 at the B locus, giving a total of 33 antigens. On these a Harpending distance matrix and an R matrix were calculated. As expected, on the plot of the first two eigenvectors of the R matrix (Fig. 1a) the paired samples (African Blacks, Japanese, Amerindians, and Europeans) clustered very closely to each other but with a considerable distance between the pairs; the American Blacks fell half way between the African and the Europeans. The antigens primarily responsible for this are identifiable from the plot of alleles on the first two eigenvectors of the S matrix. It appears that the position of the Africans is essentially influenced by A30, Bw42, Aw43 and B17, the position of the Europeans by A1 and A3, B8 and B18, the Japanese by A9 and A31, and the Amerindians by B5, B40 and B15.

A similar procedure was carried out for the C locus using eight antigens. Although this time the two European samples appeared close together, and so did the two Japanese and the Africans, there was an enormous difference between the two Amerindian samples, the separation being due to the presence of Cw1 and Cw3 in one and not in the other. The DR data, using ten antigens, also showed considerable differences between the paired samples, especially the Africans where it was due to the differing frequencies of DR8, 9 and 10. The relatively small number of Cw and DR antigens included in these analyses as well as technical difficulties in typing may be partly responsible. But it was

Figure 1: Plot of the first and second eigenvectors of the R matrix

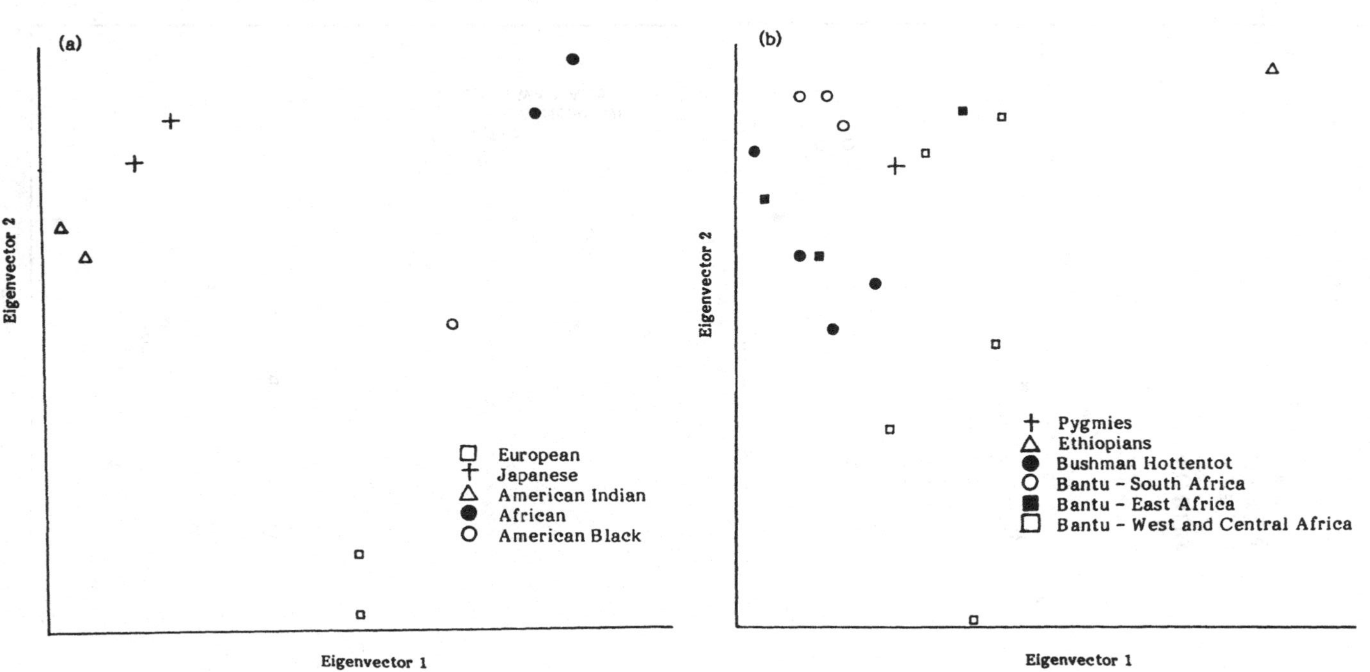

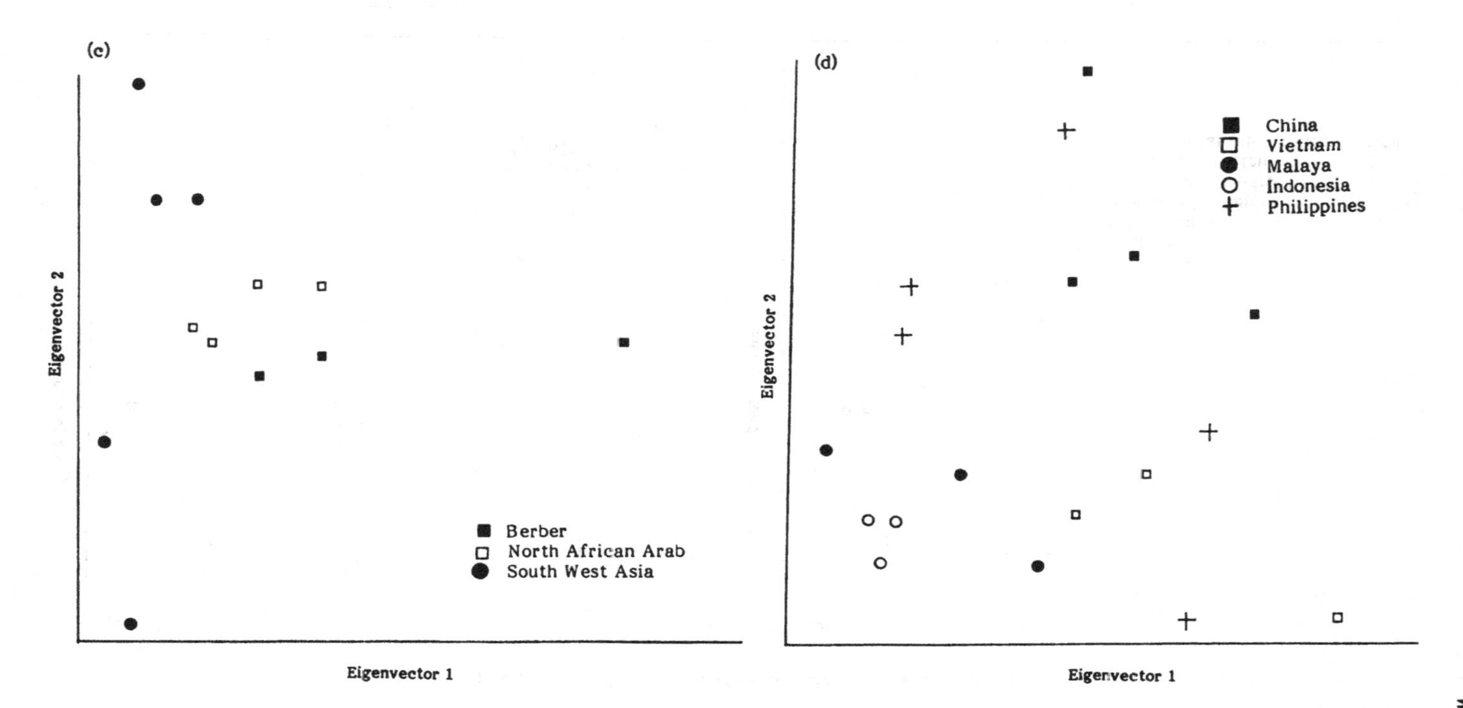
(c)
Eigenvector 2
Berber
North African Arab
South West Asia
Eigenvector 1
(d)
Eigenvector 2
China
Vietnam
Malaya
Indonesia
Philippines
Eigenvector 1

obviously necessary to examine further the variation within these continental groups, and this examination was restricted to tropical populations and to the HLA-A and B loci since data at the others were insufficient.

(b) *Intracontinental*

Turning to the tropical continents, in Africa (Fig. 1b) the Ethiopians are quite distinct from the remainder. The Bushmen Hottentot group (D to G) appears quite compact. By contrast the Bantu are quite dispersed; from South Africa samples of Zulu, Xosa, and a general sample from Durban cluster together, some East African samples appear to have been influenced by the Bushman-Hottentot group, but the samples from Zaire and of Yoruba and Sarakole in the west are very different. The Arab populations of North Africa form a fairly tight cluster, distinct from the Berbers and Tuareg and intermediate between them and populations of south-west Asia (Fig. 1c).

In south-east Asia, while the several Philippine samples are widely dispersed, the other samples seem to fall into fairly compact and distinct clusters - Indonesia, Malaya, Vietnam and China (Fig. 1d). In North America the Indians of the south-west deserts form a cluster distinct from those of Central America, while the South American tribes are widely dispersed. Similar dispersion is seen in Oceania and India where there is little overlap between north and south. Clearly there is very considerable variation among populations within continents which is masked by pooling data.

DISCUSSION

Study of the MHC region in remote populations presents particular difficulties, and the most satisfactory results require either testing in the field, or separation and immediate freezing of specimens for transmission to the laboratory; the results cannot therefore be guaranteed as free from some testing errors. Different laboratories use different batteries of antisera. Some of the samples are small and the sampling frame may not always have been as well planned as desirable. However, the data collated for the present analysis appear adequate.

Besides the well-known intercontinental differences, the results demonstrate the extent of intracontinental variation. There is in each continental area a clear tendency for clustering of populations to occur, generally along ethnic lines but with some regions showing particularly high variability, e.g. South America, the Philippines. To these clusters and the differences among them, a range of alleles make major contributions. The position of the Ethiopians relative to other Africans is predominantly influenced by the frequencies of B21, the Bushman/Hottentot by A30, B15; Chinese relative to other Orientals by A2, A11, B40; the south-west American Indians by A2, B27 and B5. Thus, with the frequencies of the HLA system, as with many other genetic systems, diversification of populations seems to have occurred in a wide variety of directions.

The MHC region shows remarkable evolutionary and phylogenetic conservatism, not only in antigen structure but also in gene organisation and function. But the biological advantage of retaining a system of such extreme complexity is not obvious. To account for its polymorphic nature, perhaps the more allotypic variants there are to choose from, the greater the chance that some members of a species would have the appropriate display of MHC molecules to mount an efficient immune response (Zinkernagel et al, 1978). The selective advantage of any particular new pathogen would be reduced, the frequency of a new allele more efficient in counteracting it would reach a dynamic equilibrium with the pathogen prevalence (Wakeland & Nadeau, 1981). If this is so, it would account for the considerable variation in antigen frequencies - different pathogens, different prevalences, and the establishment of different mutant alleles in the populations of different environments.

REFERENCES

Auffray, C., Lillie, J.W., Arnot, D., Grossberger, D., Kappes, D., & Strominger, J.L. (1984). Isotypic and allotypic variation of the human MHC class II antigen alpha-chain genes. Nature, **308**, 327.

Baur M.P. & Danilovs J.A. (1980). Population analysis of HLA-A, B, C, DR and other genetic markers. In: P.I. Terasaki (ed.), Histocompatibility Testing 1980, p. 955. Copenhagen: Munksgaard.

De Kretser, T.A., Crumpton, M.J., Bodmer, J.F. & Bodmer, W.F. (1983). Demonstration of two distinct light chains in HLA-DR associated antigens by two-dimensional gel electrophoresis. European Journal of Immunology, **12**, 214.

Duquesnoy, R.J., Marrari, M. & Annen, K. (1979). Identification of an HLA-DR associated system of B cell alloantigens. Transplantation Proceedings **11**, 1757.

Harpending, H. & Jenkins, T. (1973). Genetic distance among southern African populations, In: M.H. Crawford & P.L. Workman (eds.), Methods and Theories of Anthropological Genetics, pp. 177-199. Albuquerque: University of New Mexico Press.

Kissmeyer-Nielson, F., Svejgaard, A. & Hauge, M. (1968). Genetics of the human HL-A transplantation system. Nature, **219**, 1116.

Kissmeyer-Nielsen, F., Sørensen, S.F., Svejgaard, A., Nielsen L. Staub. (1969). Crossing over within the HL-A system. Nature, **224**, 75.

Kratzin, H., Yang, C., Gotz, H., Pauly, E., Korbel, S., Egert, G., Thinnes, F.P., Wernet, P., Altervogt, P. & Hilschmann, N. (1981). Primary structure of Class II human histocompatibility antigens. Hoppe-Seyler's Zeitschrift fur Physiologie und Chemie, **362**, 1665.

Low, B., Messeter, L., Mansson, S. & Lindholm, T. (1974). Crossing over between the SD-2 (FOUR) and SD-3 (AJ) loci of the human major histocompatibility chromosomal region. Tissue Antigens, **4**, 405.

Mach, B., Gorksi, J., Strubin, M., Long, E.O. & Depreval, C. (1984). Molecular genetics of the human class II antigens. Journal of Cell Biochemistry, 8a (suppl), 127.

Mawas, C., Charmot, D. & Mercier, P. (1980). Split of HLA-D into two regions alpha and beta by a recombination between HLA-D and GLO I. Study in a family and primed lymphocyte typing for determinants coded by the beta region. Tissue Antigens, **15**, 458

Pickbourne, P., Piazza, A. & Bodmer, W.P. (1977). Joint Report 6, Population Analysis, In: W.F. Bodmer, J.R. Batchelor, J.G. Bodmer, H. Festenstein & P.J. Morris (eds.), Histocompatibility Testing 1977, pp. 259-278. Copenhagen: Munksgaard.

Sandberg, L. Thorsby, E., Kissmeyer-Nielsen, F., & Lindholm, A. (1970). Evidence of a third sublocus within the HL-A chromosomal region. In: P.I. Terasaki (ed.), Histocompatiblity Testing 1970, p. 165. Copenhagen: Munksgaard.

Shaw, S., Johnson, A.H. & Shearer, G.H. (1980). Evidence for a new segregating series of B cell antigens which are encoded in the HLA-D region and stimulate allogeneic proliferative and cytotoxic responses. Journal of Experimental Medicine, **152**, 565.

Shaw, S., De Mars, R., Schlossman, S.P., Smith, P.L., Lampson, L.A. & Nadler, L.H. (1982). Serological identification of the human secondary B cell antigens. Correlations between function, genetics and structure. Journal of Experimental Medicine, **156**, 731.

Spelmann, R.S., Lee, J., Bodmer, W.F., Bodmer, J.G. & Trowsdale, J. (1984). Six HLA-D region alpha-chain genes on human chromosome 6: Polymorphisms and associations of DC_{alpha}-related sequences with DR types. Proceedings of the National Academy of Sciences, **81**, 3461.

Wakeland, E.K. & Nadeau, J.H. (1981). Immune responsiveness and polymorphism of the major histocompatibility complex: An interpretation. In: E. Sercarz & A.J. Cunningham (eds.), Strategies of Immune Regulation, p. 149. New York: Academic Press.

Walford, R.L., Smith, G.S., Zeller, E. & Wilkinson, J. (1975). A new alloantigenic system on human lymphocytes. Tissue Antigens, **5**, 196.

Zinkernagel, R.M., Callahan, N., Althage, A, Cooper, S., Klein, P.A., & Klein J. (1978). On the thymus in the differentiation of 'H-2 self-recognition' by T cells: evidence for dual recognition? Journal of Experimental Medicine, **147**, 882.

CHROMOSOME POLYMORPHISM IN HUMANS

H. G. SCHWARZACHER

Histologisch-Embryologisches Institut der Universität Wien, Wien, Austria

INTRODUCTION

Human chromosomes exhibit structural variants which occur at considerable frequencies. Although at the Paris Conference on Standardization in Human Cytogenetics (1975) the term "chromosomal heteromorphism" was used for the description of these variants, the term "chromosomal polymorphism" is here used, "chromosomal" to indicate that the variant is at the level of microscopically detectable chromosome structures, "polymorphism" to indicate that it complies with the conventional definition of a genetic polymorphism.

Chromosomal polymorphic variants are constant within a given individual. At least, no consistent differences between the different cell types of one individual are known (though there are special problems with the nucleolar organizer regions, see below). Chromosomal polymorphisms are inherited in a simple Mendelian mode. They can be demonstrated by several techniques, but not all of them by any one technique. They are usually not present in an "all or none" fashion but show different grades, and this makes their evaluation rather difficult in some cases.

The chromosomal polymorphisms were reviewed a few years ago by Verma and Dosik (1980). They are of course also relevant clinically, e.g. as a contributing cause of chromosome abnormalities, as possibly associated with certain diseases, and with tumorgenesis (see, e.g., Atkin, 1977; Kivi & Mikelsaar, 1980; Sutherland, 1983; Glover et al, 1984). In this paper some considerations and new findings on the nature of the chromosomal polymorphisms are presented as well as a brief review of the more recent results of population studies.

TYPES OF POLYMORPHIC REGIONS OF CHROMOSOMES

Table 1 lists different types of chromosomal polymorphisms. The different chromosome bands are placed in one group because of their cytological

Table 1: Polymorphic regions in chromosomes

Chromosome bands:
Constitutive heterochromatin or heterochromatin-like regions containing highly repetitive DNA.

C-bands: classical constitutive heterochromatin.
Fine bands: Q-, G-, R-, and special bands.
Lateral asymmetry.

Fragile sites

Major structural variants:
Inversions

Nucleolus organiser regions (NOR)

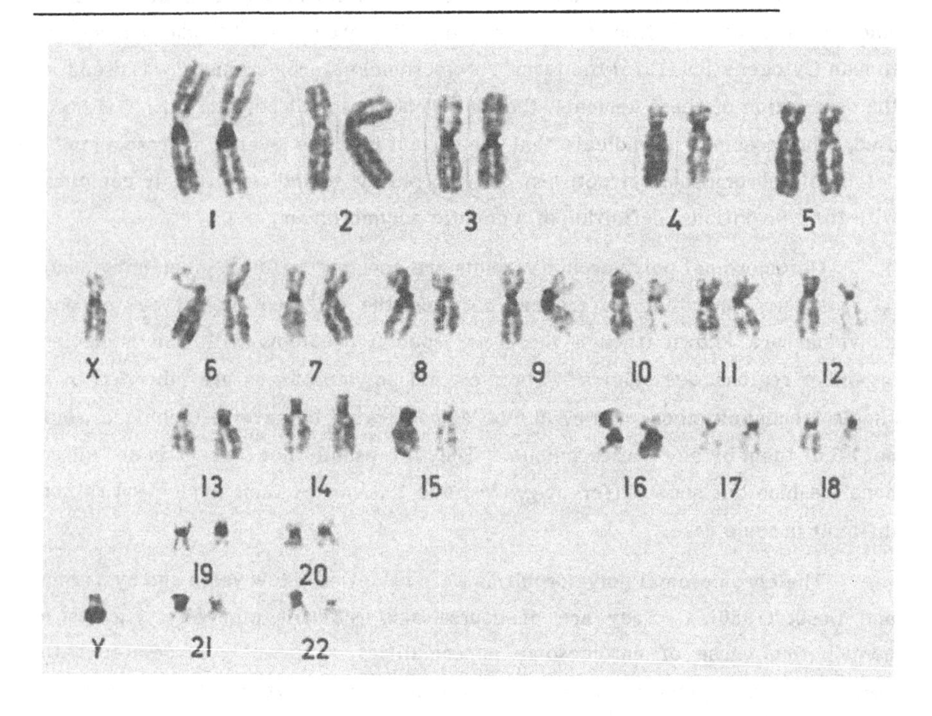

Figure 1: Human male karyotype, staining for C-bands. Large polymorphic regions in chromosomes 1, 9, 16 and Y. Clear heterozygosity is shown in chromosomes 1 and 9 in this case.

and molecular similarities. The so-called fragile sites may be regarded as polymorphic only under certain circumstances. Among the rearrangements of large chromosome regions probably only inversions are found frequently enough to justify their classification as a polymorphism. The nucleolar organizer regions (NOR) show a clear polymorphism but the interpretation of cytological methods demonstrating NOR-activity requires caution.

Chromosome bands

A chromosome band is defined as a region of a chromosome (or chromatid) which can be clearly discerned from its neighbouring region by any microscopic technique. A certain number of bands are polymorphic. These all have in common the fact that they contain highly repetitive DNA. Since the amount and the classes of repetitive DNAs vary in the different types of polymorphic bands, they react differently to various cytological techniques. Also differences in the packing of the DNA-protein fibril (the elementary chromosome fibril) and differences in the protein components may play a role in differential staining (reviewed by Schwarzacher, 1976; Jack et al, 1985). The differing amounts of repetitive DNA may explain why a given band may vary in size, to produce the pattern of continuous variation that is seen: the polymorphism can be caused by the presence of more or fewer copies of the redundant DNA.

Repetitive DNA sequences, with the exception of NORs, are presumably not translated. At least, constitutive heterochromatic regions must be regarded as genetically inert. On the one hand this inactivity may account for their polymorphism - one that is neutral; on the other hand the fact that they are found in the genomes of practically all eukaryotic species suggests that they may have important functions, albeit unknown - a functional polymorphism.

C-bands

C-bands are composed of constitutive heterochromatin. The staining procedure for their identification was introduced by Arrighi and Hsu (1971). In the human genome (Fig. 1) constitutive heterochromatin is found in all chromosomes in the centromeric regions. Larger blocks are situated in the secondary constrictions of chromosomes 1, 9 and 16, and in the distal part of the long arm of the Y-chromosome. The larger C-regions are clearly polymorphic. In Figure 1 chromosomes 1 and 9 show distinct differences in size and staining intensities of the C-bands in the two homologues. The polymorphism demonstrated by the C-band staining has also been shown by *in situ* hybridisation of satellite-cRNAs to chromosome preparations (e.g. Gosden et al, 1981) and can be observed in electron microscope preparations (Harrison et al, 1983, 1985).

Although the size of the polymorphic C-bands shows a gradual and continuous variation, most earlier investigators tried to classify it into groups. Patil and Lubs (1977), for example, used the size of the short arm of chromosome 16 as a standard, and set up five size classes. Absolute measurements of the size of the large C-bands were introduced by Podugolnikova et al (1979) and Erdtmann et al (1981). These authors set C-band sizes in relation to the length of chromosome 1 or to the total C-band length, in order to correct for chromosome contraction differences or to obtain relative values for single C-bands. These methods are preferable by far to simple estimates of size and subjective classifications.

Q-bands

The introduction of the fluorochrome quinacrine (Caspersson et al, 1968; Zech, 1969) revealed for the first time a fine banding pattern of the chromosomes and shortly thereafter the first Q-polymorphisms were described (reviewed by Verma & Dosik, 1980). So-called "prominent" Q-bands (see Fig. 2) are found in the centromeric regions of chromosomes 3 and 4, at two sites of the short arms (p11 and p13) of all acrocentric chromosomes, and in the long arm of the Y-chromosome. All these prominent Q-bands show a polymorphism which is expressed in weaker or stronger fluorescence and in different sizes of the bands. It must be emphasized that only some of the C-bands show a strong fluorescence with quinacrine, others not (e.g., in chromosome 1). This depends on the A-T-content, since quinacrine stains preferentially A-T-rich chromosome regions (Weisblum & De Haseth, 1972).

The evaluation of Q-bands is difficult because a subjective estimation has to be made in most cases. Photometric measurements (e.g. Van der Ploeg et al, 1974; Schnedl et al, 1977b) brought no real improvement, because the main variability due to the artefacts is caused by the quality of the cytological preparation and the fluorescence stain. A definite improvement of the latter was the introduction of nonfading fluorochromes (DAPI and DIPI, see Schnedl et al, 1977a).

The Paris Conference (1971) recommended classification of the fluorescence intensity into five grades, whereby each grade is compared with a particular Q-band in the same metaphase. In this way a more reliable relative

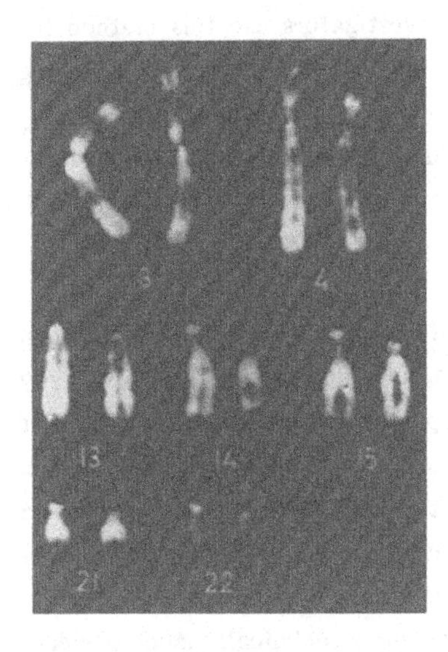

Figure 2: Q-banding of polymorphic chromosomes 3, 4 and the acrocentrics of a female, all showing heterozygosity.

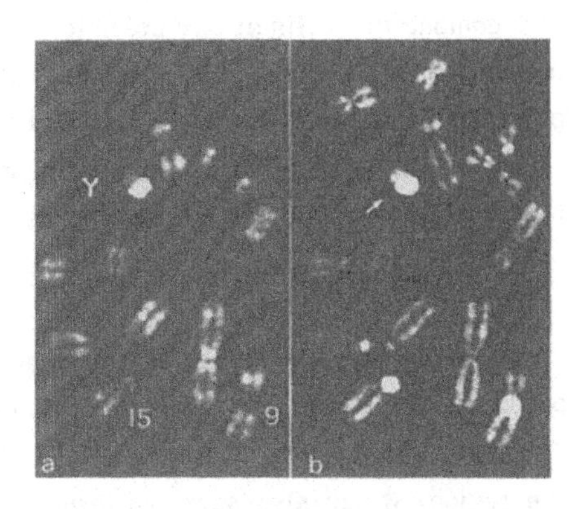

Figure 3: Part of a metaphase of a normal male. (a) Q-banding. (b) Staining with D178/170 showing large fluorescent blocks in chromosomes 9 and 15; also stronger fluorescence of Y and of the short arms of the acrocentric chromosomes. (From Schnedl et al, 1981.)

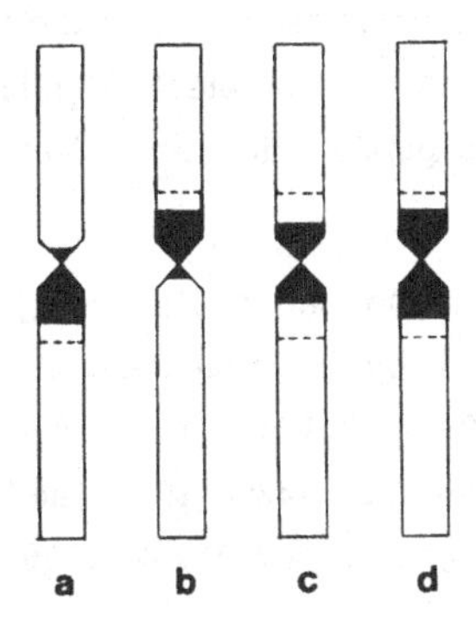

Figure 4: Schematic drawing of variants of the centromeric C-band in chromosomes 1, 9 and 16. (a) Most frequent (over 90%), size of band in the long arm variable. (b) Interpretation as inversion possible. (c),(d) Different sizes of the band in the short arm as well as in the long arm, usually also called "inversion".

and comparable estimate is possible. Most investigators use this method for population studies. To illustrate the value of Q-banding in studying polymorphism, the 7 chromosome pairs bearing prominent Q-bands of a DAPI-stained cell of a female are shown in Figure 2. Each of them shows heterozygosity of the polymorphic bands.

G-bands

A number of different treatments of chromosome preparations can produce a fine banding pattern when Giemsa-staining is performed (see review by Schnedl, 1974). With these methods polymorphisms are detectable in most of the regions which also show variations by the C- and Q-banding techniques. Particularly the secondary constrictions and the satellites of the acrocentric chromosomes can be clearly demonstrated. Nevertheless, G-banding techniques have not so far been used for more extensive studies of chromosomal polymorphism, presumably because fluorescence-staining is technically much easier.

Special banding methods

There are several ways of increasing the accuracy as well as the resolution of the banding techniques. One is to make use of the slender prophase chromosomes which show little contraction. Schnedl (1971) already in his first paper on G-bands reported that about 250 bands per haploid set are observable in early metaphase chromosomes with the light microscope, and Bahr et al (1973) counted about 500 in the electron microscope. Later studies enabled Yunis et al (1978) to publish diagrams of the very fine prophase bands on the chromosomes.

Another method is the so called R-banding (reverse-banding). The original method was introduced by Dutrillaux and Lejeune (1973). The use of acridine orange to produce R-bands (RFA-bands, reverse fluorescence acridine orange) was modified by Verma and Lubs (1975), and with this method further polymorphisms can be found (see Verma & Dosik, 1980).

Finally, there is the use of A-T- and G-C-specific fluorochromes individually, in combination with each other, and in combination with counterstaining A-T- or G-C-specific agents to enhance or suppress the fluorescence (Schweizer, 1981). For instance, six polymorphic bands can be discerned in the short arm of chromosome 15 with such methods (Wachtler & Musil, 1980), and Buys et al (1981) were able to demonstrate two different polymorphic sites within the C-band of chromosome 9 using a combined distamycin-DAPI staining. In

Figure 3, part of a metaphase stained with the fluorochrome D 178/170 (Schnedl et al, 1981) is shown in comparison with quinacrine staining. This fluorochrome stains short arms and satellites of the acrocentric chromosomes better than quinacrine and a particularly bright fluorescence is seen in chromosomes 15 and 9.

Lateral asymmetry

"Lateral asymmetry" refers to an unequal distribution of the two bases of a base pair in the two strands of the DNA molecule. In the case of A-T this can be made visible by *in vitro* treatment of cells with bromodeoxyuridine (BrdU). BrdU binds preferably to thymine during DNA replication. In the chromosomes of the following metaphase the T-sites are therefore occupied by BrdU (instead of A) which prevents their staining with A-T-specific fluorochromes such as quinacrine or DAPI, whereas the same regions of the sister chromatids (containing A-T) are brightly stained (Lin et al, 1974). Lateral asymmetry has been described for almost all polymorphic Q-bands and C-bands (review by Brito-Babapulle, 1981). A distinct polymorphism of lateral asymmetry in chromosome 1 was demonstrated by Lin and Alfi (1978), but no extensive population studies have been published so far.

Fragile sites

A "fragile site" is seen as a distinct non-staining gap (using standard orcein or Giemsa staining) usually in both chromatids of a metaphase chromosome. As a result of breaks at these sites, acentric fragments, deleted chromosomes, and multiradial chromosome figures may develop. The expression of fragile sites, i.e. their visibility in chromosome preparations of lymphocyte cultures, is greatly enhanced by specific variations of the culture conditions. As a matter of fact, the fragile site on chromosome X was only discovered because the cells were grown by chance in a medium deficient in folic acid and thymidine. The following factors can lead to an expression of fragile sites (Sutherland et al, 1985): deficiency of folic acid; distamycin A; aphidicolin; 5-azacytidine; BrdU.

Of the 29 fragile sites known so far, 18 are considered to be very rare. They are all inherited in a Mendelian fashion. The rare fragile site in the X-chromosome is associated with a distinct pathological condition and was not found among 1019 newborns with normal phenotype (Sutherland, 1983). The remaining 11 fragile sites are called "common" (Glover et al, 1984; Sutherland et al, 1985). Their frequencies in the population seem to be quite high but in most cases are not exactly known. A definite polymorphism has been reported for the fragile site 10q25 (BrdU inducible) in the Australian population (Sutherland, 1983). Further population studies may well prove that other fragile sites are also polymorphic.

Major structural variants

Usually structural variants of the chromosomes in which larger parts (larger than a "band") are involved, are infrequent and considered as abnormalities. The only known major structural arrangements found in an appreciable proportion of phenotypically normal persons are inversions of the centric regions in chromosomes 1, 9 and 16. It is, however, not clear whether all "inversions" reported in the literature are true inversions, or just extensions of the centromeric bands. It would be a true inversion if the centromeric band, which in the majority of cases extends farther into the long arm, is seen to extend into the short arm instead, with the long arm showing only a small band (Fig. 4b). In some cases, however, the heterochromatic band is found in both arms to variable degrees (Figs. 4c,d). All these variants are nevertheless usually called "inversions" (e.g., Soudek & Stroka, 1978; Verma et al, 1981).

Nucleolus organizer regions

The term nucleolus organizer region (NOR) is used for a chromosome region which contains the genes for 18S- and 28S-rRNA. In man, NORs are located on the short arms of all acrocentric chromosomes, and there form characteristic secondary constrictions. Each NOR contains usually many rRNA gene copies.

The polymorphism of the NORs is twofold: 1. There are individual differences in the number of rRNA genes. Therefore the sizes of NORs may differ and one or several NORs may even be totally missing, or at least not detectable by cytological methods. 2. The activity of the rRNA-genes may vary. This is of course dependent on the protein synthesis activity of the cell requiring ribosomes.

Several staining procedures for acidic proteins are able to stain differentially the NORs in hypotonic pretreated chromosome preparations. The most common is the silver staining introduced by Howell et al (1975) and Goodpasture and Bloom (1975). These techniques make visible only those NORs that were transcribed during the preceding interphase, since they stain the proteins around

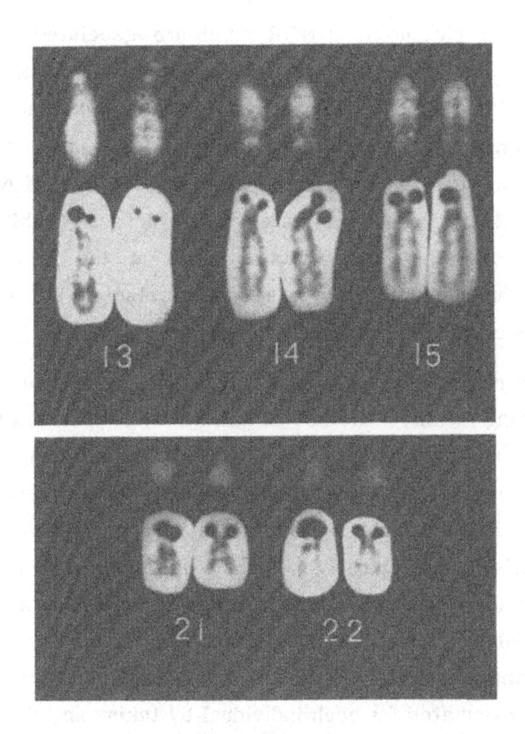

Figure 5: The five pairs of NOR-bearing chromosomes. Upper rows: Q-banding. Lower rows: silver staining. Despite differences in the intensity of silver staining, all ten chromosomes should be registered in this case as "positive".

Table 2: Modal number of silver positive NORs ($Ag^{+}NOR$) in different populations. N = Number of individuals investigated.

Population	N	$Ag^{+}NOR$	Authors
Austrian and German (Vienna and Ulm)	51	8.7	Mikelsaar et al, 1977
Estonian	41	7.8	Mikelsaar & Ilus, 1979
Russian (Moscow)	40	8.4	Zakharov et al, 1982
French	100	8.58	Dipierri & Fraisse, 1983
East Indians	70	8.05	Verma et al, 1981

and attached to the chromosomal NOR which are associated with transcription or with pre-ribosomal structures (review by Schwarzacher & Wachtler, 1983; Goessens, 1984; Hügle et al, 1985). If metaphase chromosomes of rapidly dividing cells, e.g. PHA-stimulated lymphocytes *in vitro*, are observed, the silver staining presumably represents the actual sizes and numbers of NORS because in such cells the rDNA is probably activated to the highest possible degree. It has in fact been shown that the NORs as seen in the *in situ* rRNA-rDNA hybridization technique are visible to a similar extent in silver preparations in such cells (Evans et al, 1974; Warburton & Henderson, 1979).

For these reasons, it is necessary to restrict population studies of NOR-polymorphism using the silver method to uniform cell types in a highly activated form. PHA-stimulated lymphocytes *in vitro* are suitable material. Since variations in size and intensity of silver deposits also depend very much on technical factors it is advisible to restrict the scoring to simply "silver-positive" and "silver-negative". In Figure 5 an example of an individual is shown, in whom all 10 NORs are silver-positive. Although there are considerable differences in the size of the silver deposits, all NORs would be counted as "positive" in this case. It is furthermore recommended that a "modal number" of silver-positive NORs should be calculated for each individual by taking an NOR which is silver-positive in more than 50% of the cells as positive (Table 2). Such a procedure may of course introduce a number of errors, but has been found useful when different populations have been compared.

POPULATION STUDIES

Many of the earlier chromosomal polymorphism studies cannot be used for comparisons between different populations because each relates to only a single population investigated by one author or one laboratory. Methods of demonstrating, classifying and evaluating chromosomal polymorphisms vary between different laboratories to such an extent that only studies in which the same standardized methods of cytological staining, analysis, and measurement are used will give reliable comparative results. Best results are of course obtained when different populations are compared by one laboratory.

Comparisons between populations have so far been carried out mainly using C- and Q- bands including the inversions in chromosomes 1, 9, and 16, and NORs. Special banding procedures have been applied on a larger scale only by Verma and Dosik (1981), but have not yet brought decided new results. Fragile sites are at present of much interest for the problem of induced chromosome breakages and tumorgenesis and there is considerable scope for their application also in population studies.

In the following, some of the more recent findings on population studies of C- and Q- bands are reported. The Y chromosome is dealt with separately because its polymorphic band can be demonstrated with either method. Finally the results on NOR-studies are reviewed. For earlier studies the papers by Müller et al (1975), Mikelsaar et al (1975), Yamada and Hasegawa (1978), and the review by Verma and Dosik (1980) should be consulted. There has been relatively little on tropical peoples, but such as there is is put into the perspective of studies outside the tropics.

C-bands

Exact measurements introduced by Podugolnikova et al (1979a) proved useful, for they revealed a random combination of the polymorphic sites within individual karyotypes and no sex differences in a group of 50 normal boys and 50 normal girls (age range 4-14 years), presumably from Moscow (Podugolnikova et al, 1979b; Podugolnikova, 1979). A reduction in overall C-band size was found in children with embryopathies of unknown etiology as well as with Down's syndrome (Podugolnikova et al, 1984).

Erdtmann et al (1981), using densitometric measurements of C-bands, found distinct differences between Caucasoids and five different tribes of Brazilian Indians. In the Indians only minor intertribal variations were observed but Caucasoids presented on average lower C-band values. This study is of real importance, being the only one (except studies on the Y chromosome) using measurements and a comparison of different populations.

Ibraimov et al (1982) compared five ethnically closely-related groups of Mongoloids of Central Asia living under different ecological conditions. These authors used the "semi-quantitative" method of comparing C-band sizes with a defined autosome (Patil & Lubs, 1977). They could not find any differences between these groups. Berger et al (1983) compared populations from Sweden and France, using also the semi-quantitative approach. They were able to find small but significant differences, the Swedish population showing slightly bigger C-bands in all sites and more "inversions" in chromosome 9.

The C-band studies seem to hint that ethnic differences may exist (e.g., Erdtmann et al, 1981), whereas ecological influences may not be so important (e.g. Ibraimov et al, 1982).

Q-bands

As already explained, the analysis of Q-bands has to rely on subjective judgment of the brilliance of the fluorescence. Al-Nassar et al (1981) reviewed the studies using the recommendations of the Paris Conference (1971) and discussed them in relation to their own findings on three different populations from Kuwait. They came to the conclusion that the studies of Lubs et al (1977) and of Müller et al (1975) show a tendency to a higher incidence of Q-bands in the American black population than in the white population. Among the Kuwait populations the Ajman showed the most similarities to the American blacks indicating a black influence in this population.

Ibraimov and co-workers (summarised by Ibraimov, 1983) showed in extensive studies that there is a high level of homogeneity in all populations with regard to the relative Q-band material per chromosome, but that differences exist in the absolute amount of Q-band material per genome. This is best demonstrated by counting the mean number of Q-bands per individual. Analysing the data from this point of view they found a trend towards a decrease of Q-material in populations living in high altitude or under extreme climatic conditions. This may overlap possible ethnic differences. By contrast to the C-bands, this may indicate that Q-band variation may be influenced by ecological factors.

Ibraimov (1983) reported also a tendency to a sex difference in all populations investigated, in that females in general have in their seven polymorphic autosomes more Q-band material than males. He discussed the possibility that this may represent compensation for the rather large Q-bands in the Y chromosomes.

Y chromosomes

The distal part of the Y chromosome is heterochromatic, thus forming a C-band, and this is at the same time the most brilliantly fluorescing Q-band. Although the C-band and the Q-band stainings may not identify exactly the same region, and although lateral asymmetry studies show two different regions (Limon et al, 1979), these can be regarded as a single region for practical reasons. The band varies to such a high degree that it causes considerable variation of the total length of the Y chromosome. Indeed already in 1966, before the introduction of banding techniques, comparative studies revealed distinct population differences in the size of the Y chromosome (Cohen et al, 1966), supported by more recent analyses (Chakraborty & Chakraborty, 1984).

A reliable method of measuring the size of the Q-band of the Y chromosome is to compare it with specified autosomes of the same metaphase. A "large" Y chromosome is scored, e.g. if it is equal to or larger than chromosome 19. The frequencies of the occurrence of large Y chromosomes show clear population differences. Al-Nassar et al (1981) summarized the findings to date, and their Table 8 suggests the following approximate order of decreasing size of the polymorphic band of the Y chromosome (Yq12): East Asians, Jews, Blacks, American Indians, Caucasians.

The possibility that the length of the heterochromatic band of the Y chromosome may be connnected with body height has also been discussed (e.g., Yamada et al, 1981; Verma & Dosik, 1982).

Nucleolus organizer regions

The difficulty of scoring "active" NORs using silver staining has been discussed above. In the present state of the problem only those studies from different laboratories can be compared if they use the "modal number" of NORs per individual. In Table 2 such studies are summarised after excluding earlier ones on limited numbers of individuals. The first two listed (Mikelsaar et al, 1977; Mikelsaar & Ilus, 1979) were made using strictly standardised techniques and allow a direct comparison. The studies by Zakharov et al (1982) and by Dipierri and Fraisse (1983) are made under quite similar conditions and may be included for comparison with only slight reservation. In the study of Verma et al (1981) on East Indians, while modal numbers of silver-positive NORs were established, possible differences in the silver staining techniques may render these values less comparable. There has been published a study on Georgians in which an NOR modal number of 9 was found (Kristesoshivili et al, 1983), but only an English summary of this paper has come to my knowledge; presumably it used the same standardized method as Zakharov et al (1982).

This leaves us with the result, that European Caucasians have a slightly higher NOR modal number than Estonians (which can be regarded as a mixed population of Ugric Finns and Caucasians). This difference has been shown to be significant between the Austria/South-Germany (Vienna/Ulm) and the Estonian (Tartu) populations (Mikelsaar & Ilus, 1979). Further population studies on NORs may therefore be promising.

CONCLUSIONS

Chromosomal polymorphisms can be studied by a series of methods. So far there are reliable studies comparing different populations with respect to

C-bands, Q-bands, and NORs. Considering the complexity and laboriousness of the methods that have to be employed, the results of these studies are not overwhelming. There is an indication that ethnic factors contribute to variation in the size of the C-bands, particularly on the Y chromosome. The autosomal Q-bands may be influenced also by ecological factors. The NORs show some indications of racial differences.

Future studies can only give reliable results if the cytogenetic and the analytical procedures are strictly standardised and made on sufficiently large samples of populations. The use of the more refined cytological methods, e.g. the fluorescence double and counter-stainings, as well as consideration of fragile sites, are recommended.

REFERENCES

Al-Nassar, K. E., Palmer, C. G., Conneally, P. M. & Yu, P. L. (1981). The genetic structure of the Kuwaiti population. II: The distribution of Q-band chromosomal heteromorphisms. Human Genetics, **57**, 423-427.

Arrighi, F. & Hsu, T. C. (1971). Localization of heterochromatin in human chromosomes. Cytogenetics, **10**, 81-86.

Atkin, N. B. (1977). Chromosome 1 heteromorphism in patients with malignant disease: a constitutional marker for a high risk group? British Medical Journal, **1**, 358.

Bahr, G. F., Mikel, U. & Engler, W. F. (1973). Correlates of chromosomal bandings at the level of ultrastructure. In: T. Caspersson & L. Zech (eds.), Nobel Symposium 23: Chromosome identification - technique and applications in biology and medicine. London/New York: Academic Press.

Berger, R., Bernheim, A., Kristoffersson, U., Mineur, A. & Mitelman, F. (1983). Differences in human C-band pattern between two European populations. Hereditas, **99**, 147-149.

Brito-Babapulle, V. (1981). Lateral asymmetry in human chromosomes 1, 3, 4, 15 and 16. Cytogenetics & Cell Genetics, **29**, 198-202.

Buys, C. H. C., Gouw, W. L., Blenkers, J. A. M. & Dalen, C. H. van (1981). Heterogeneity of human chromosome 9 constitutive heterochromatin as revealed by sequential distamycin A/DAPI staining and C-banding. Human Genetics, **57**, 28-30.

Caspersson, T., Farber, S., Foley, G. E., Kudynowski, J., Modest, E. J., Simonsson, E., Wagh, U. & Zech, L. (1968). Chemical differentiation along metaphase chromosomes. Experimental Cell Research, **49**, 218-222.

Cohen, M., Shaw, M. W. & MacCluer, J. W. (1966). Racial differences in the length of the human Y chromosome. Cytogenetics, **5**, 34-52.

Dipierri, J. E. & Fraisse, J. (1983). Polymorphism des bandes NOR dans une population francaise. Annales Genetiques (Paris), **26**, 215-219.

Dutrillaux, B. & Lejeune, J. (1971). Sur une nouvelle technique d'analyse du caryotype humain. C.R.Acad.Sci.(Paris), **272**, 2638-2640.

Erdtmann, B., Salzano, F. M., Mattevi, M. S. & Flores, R. Z. (1981). Quantitative analysis of C bands in chromosomes 1, 9 and 26 of Brazilian Indians and caucasoids. Human Genetics, **57**, 58-63.

Evans, H. J., Buckland, R. A. & Pardue, M. L. (1974). Location of genes coding for 18S an 28S ribosomal RNA in the human genome. Chromosoma, **48**, 405-426.

Glover, T. W., Berger, C., Coyle, J. & Echo, B. (1984). DNA polymerase and inhibition by aphidicolin induces gaps and breaks at common fragile sites in human chromosomes. Human Genetics, **67**, 138-142.

Goessens, G. (1984). Nucleolar structure. International Reviews in Cytology, **87**, 107-158.

Goodpasture, C. & Bloom, S. E. (1975). Visualization of nucleolar organizer regions in mammalian chromosomes using silver staining. Chromosoma **53**, 37-50.

Gosden, J. R., Lawrie, S. S. & Cooke, H. J. (1981). A cloned repeated DNA sequence in human chromosome heteromorphisms. Cytogenetic & Cell Genetics, **29**, 32-39.

Harrison, C. J., Allen, T. D. & Harris, R. (1983). Scanning electron microscopy of variations in human metaphase chromosome structure revealed by Giemsa banding. Cytogenetics and Cell Genetics, **35**, 21-27.

Harrison, C. J., Jack, E. M., Allen, T. D. & Harris, R. (1985). Investigation of human chromosome polymorphisms by scanning electron microscopy. Journal of Medical Genetics, **22**, 16-23.

Howell, W. M., Denton, T. E. & Diamond, J. R. (1975). Differential staining of the satellite regions of human acrocentric chromosomes. Experentia, **31**, 260-262.

Hügle, B., Hazan, R., Scheer, U., Franke, W. W. (1985). Localization of ribosomal protein S1 in the granular component of the interphase nucleolus and its distribution during mitosis. Journal of Cell Biology, **100**, 873-886.

Ibraimov, A. I. (1983). Human chromosomal polymorphism. VII: The distribution of chromosomal Q-polymorphic bands in different human populations. Human Genetics, **63**, 384-391.

Ibraimov, A. I., Mirrakhimov, M. M., Nazarenko, S. A. & Axenrod, E. I. (1982). Human chromosomal polymorphism. II: Chromosomal C polymorphism in mongoloid populations of central Asia. Human Genetics, **50**, 8-9.

Jack, E. M., Harrison, C. J., Allen, T. D. & Harris, R. (1985). The structural basis for C-banding. A scanning electron microscopy study. Chromosoma **91**, 363-368.

Kivi, S. & Mikelsaar, A. V. (1980). Q- and C-band polymorphisms in patients with ovarian or breast carcinoma. Human Genetics, **56**, 111-114.

Kristesashvili, D. I., Benjush, V. A., Egolina, N. A. & Davudov, A. Z. (1983). Comparative analysis of the polymorphism of human silver staining chromosomal nucleolus organizer regions (Russian). Genetika (Moskva), **19**, 1205-1209.

Limon, J., Gibas, Z., Kaluzewski, B. & Moruzgala, T. (1979). Demonstration of two different regions of lateral asymmetry in human Y chromosomes. Human Genetics, **51**, 247-252.

Lin, M. S. & Alfi, O. S. (1978). Variation in lateral asymmetry of human chromosome 1. Cytogenetics and Cell Genetics, **21**, 243-250.

Lin, M. S., Latt, S. A. & Davidson, R. L. (1974). Microfluorometric detection of asymmetry in the centromeric region of mouse chromosomes. Experimental Cell Research, **86**, 392-394.

Lubs, H. A., Patil, S. A., Kimberling, W. J., Brown, J., Cohen, M., Gerald, P., Hecht, F., Myrianthopoulos, N. & Summitt, R. L. (1977). Q- and C-band polymorphisms in 7- and 8-year-old children: racial differences and clinical significance. In: E. B. Hook & I. H. Porter (eds.), Population cytogenetics: studies in humans, pp. 133-159. New York: Academic Press.

Mikelsaar, A. V. & Ilus, T. (1979). Population polymorphisms in silver staining of nucleolus organizer regions in human acrocentric chromosomes. Human Genetics, **51**, 281-285.

Mikelsaar, A. V., Kaosaar, M. E., Tuur, S. J., Viikmaa, M. H., Talvik, T. A. & Laats, J. (1975). Human karyotype polymorphism. III: Routine and fluorescent microscopic investigation of chromosomes in normal adults and mentally retarded children. Human Genetics, **26**, 1-23.

Mikelsaar, A. V., Schmid, M., Krone, W., Schwarzacher, H. G. & Schnedl, W. (1977). Frequency of Ag-stained nucleolus organizer regions in the acrocentric chromosomes of man. Human Genetics, **37**, 73-77.

Müller, H., Klinger, H. P. & Glasser, M. (1975). Chromosome polymorphism in a human newborn population. II: Potentials of polymorphic variants for characterizing the idiogram of an individual. Cytogenetics and Cell Genetics, **15**, 239-255.

Paris Conference on Standardization in Human Cytogenetics (1971). Birth defects. Original Articles Series VIII. New York: The National Foundation.

Paris Conference on Standardization in Human Cytogenetics (1975). Birth defects. Original Articles Series XI. New York: The National Foundation.

Patil, S. R. & Lubs, H. A. (1977). Classification of qh regions in human chromosomes 1, 9, and 16 by C-banding. Human Genetics, **38**, 35-38.

Ploeg, M. van den, Duijn, P. van & Ploem, J. S. (1974). High resolution scanning densitometry of photographic negatives of human metaphase chromosomes. I: Instrumentation. Histochemistry **42**, 9-29.

Podugolnikova, O. A. (1979). The quantitative analysis of polymorphism on human chromosomes 1, 9, 16, and Y. III: Study of relationships of C-segments length in individual karyotypes. Human Genetics, **49**, 261-268.

Podugolnikova, O.A., Parfenova, I.V., Sushanlo, H. M., Prokofieva-Belgovskaja, A. A. (1979a). The quantitative analysis of polymorphism on human chromosomes 1, 9, 16, and Y. I: Description of individual karyotypes. Human Genetics, **49**, 243-250.

Podugolnikova, O. A., Sushanlo, H. M., Parfenova, I. V., Prokofieva-Belgovskaja, A. A. (1979b). The quantitative analysis of polymorphism on human chromosomes 1, 9, 16, and Y. II: Comparison of the C segments in male and female individuals (group characteristics). Human Genetics, **49**, 251-260.

Podugolnikova, O. A., Grigorjeva, N. M. & Blumina, M. G. (1984). Relationship of the variability of the heterochromatic regions of chromosomes 1, 9, 16, and Y to some anthropometric characteristics in children with embryopathies of unknown etiology and in children with Down's syndrome. Human Genetics, **68**, 254-257.

Schnedl, W. (1971). Analysis of the human karyotype using a re-association technique. Chromosoma **34**, 448-454.

Schnedl, W. (1974). Banding patterns in chromosomes. International Review of Cytology (Suppl.), **4**, 237-272.

Schnedl, W., Mikelsaar, A. V., Breitenbach, M. & Dann, O. (1977a). DIPI and DAPI: fluorescence banding with only negligible fading. Human Genetics, **36**, 167-172.

Schnedl, W., Roscher, U. & Czaker, R. (1977b). A photometric method for quantifying the polymorphisms in human acrocentric chromosomes. Human Genetics, **35**, 185-191.

Schnedl, W., Abraham, R., Dann, O., Geber, G. & Schweizer, D. (1981). Preferential fluorescent staining of heterochromatic regions in human chromosomes 9, 15 and the Y, by D287/170. Human Genetics, **59**, 10-13.

Schwarzacher, H. G. (1976). Chromosomes in mitosis and interphase. In: W. Bargmann (ed.), Handbuch der mikroskopischen Anatomie des Menschen I/3. Berlin/Heidelberg/New York: Springer-Verlag.

Schwarzacher, H. G. & Wachtler, F. (1983). Nucleolus organizer regions and nucleoli. Human Genetics, **63**, 89-99.

Schweizer, D. (1981). Counterstain-enhanced chromosome banding. Human Genetics, **57**, 1-14.

Soudek, D. & Stroka, H. (1978). Inversion of "fluorescent" segment in chromosome 3: a polymorphic trait. Human Genetics, **44**, 109-115.

Sutherland, G. R. (1983). The fragile X chromosome. International Review of Cytology, **81**, 107-143.

Sutherland, G. R., Parslow, M. I. & Baker, E. (1985). New classes of common fragile sites induced by 5-azacytidine and bromdeoxyuridine. Human Genetics, **69**, 233-237.

Verma, R. S. & Dosik, H. (1980). Human chromosomal heteromorphisms: nature and clinical significance. International Review of Cytology, **62**, 361-383.

Verma, R. S. & Dosik, H. (1981). Human chromosomal heteromorphisms in American blacks. III: Evidence for racial differences in RFA color and QFQ intensity heteromorphisms. Human Genetics, **56**, 329-337.

Verma, R. S. & Dosik, H. (1982). Human chromosomal heteromorphisms in American blacks. VII: Correlation (r) between height and size of the Y chromosome. Japanese Journal of Human Genetics, **27**, 43-46.

Verma, R. S. & Lubs, H. A. (1975). A simple banding technique. American Journal of Human Genetics, **27**, 110-117.

Verma, R. S., Rodriguez, J. & Dosik, H. (1981a). Human chromosome heteromorphisms in American blacks. II: Higher incidence of pericentric inversions of secondary constriction regions. American Journal of Medical Genetics, **8**, 17-25.

Verma, R. S., Benjamin, C., Rodriguez, J. & Dosik, H. (1981b). Population heteromorphisms of Ag-stained nucleolus organizer regions (NORs) in the acrocentric chromosomes of East Indians. Human Genetics, **59**, 412-415.

Wachtler, F. & Musil, R. (1980). On the structure and polymorphism of the human chromosome no. 15. Human Genetics, **56**, 115-118.

Warburton, D. & Henderson, A. S. (1979). Sequential silver staining and hybridization in situ on nucleolus organizing regions in human cells. Cytogenetics & Cell Genetics, **24**, 168-175.

Weisblum, B. & De Haseth, P. L. (1972). Quinacrine, a chromosome stain specific for deoxyadenylate-deoxythymidylate-rich regions in DNA. Proceedings of the National Academy of Sciences (Washington), **69**, 629-632.

Yamada, K. & Hasegawa, T. (1978). Types and frequencies of Q-variant chromosomes in a Japanese population. Human Genetics, **44**, 89-98.

Yamada, K., Ohta, M., Yoshimura, K. & Hasekura, H. (1981). A possible association of Y chromosome heterochromatin with stature. Human Genetics, **58**, 268-270.

Yunis, J. J., Sawyer, J. R. & Ball, D. W. (1978). The characterisation of high-resolution G-banded chromosomes of man. Chromosoma, **67**, 293-307.

Zakharov, A. F., Davudov, A. Z., Benjush, V. A. & Egolina, N. A. (1982). Polymorphisms of Ag-stained nucleolar organizer regions in man. Human Genetics, **60**, 334-339.

Zech, L. (1969). Investigation of metaphase chromosomes with DNA-binding fluorochromes. Experimental Cell Research, **58**, 463.

RESTRICTION FRAGMENT LENGTH POLYMORPHISMS IN THE HUMAN GENOME

D.N COOPER[1] and J. SCHMIDTKE[2]

[1]*Neurochemistry Department, Institute of Neurology, London, U.K.*
[2]*Institut für Humangenetik der Universität Gottingen, Gottingen, Federal Republic of Germany.*

INTRODUCTION

Restriction fragment length polymorphisms (RFLPs) result from DNA base-pair changes that introduce or remove a restriction site, or sequence deletions, additions or rearrangments, that affect the length of DNA between sites. RFLPs promise to be useful in a number of different ways. Their utilisation as 'genetic signposts' should eventually permit the saturation of the human genome with evenly-spaced marker loci (White et al, 1985). This new system of markers at the DNA level promises to unify mapping at the cytogenetic and molecular levels, greatly improve the resolution so necessary for accurate gene mapping and linkage studies and give a new perspective on genetic variation. The improved linkage map should provide sufficient markers to localize and diagnose many hitherto undetectable genetic defects and allow the identification of heterozygous carriers.

This article describes a search for RFLPs in the human genome using a random sample of cloned DNA segments. Analysis of the data has permitted a first estimate of heterozygosity in the human genome, an amount large enough to demonstrate the extensive variation which can be exploited in clinical medicine. The clinical applications of recombinant DNA technology to the analysis and diagnosis of human genetic disease are then presented. RFLPs associated either with gene regions or linked DNA segments may permit antenatal diagnosis in cases where it has not proved possible to detect gene defects directly using a cloned gene probe.

DNA POLYMORPHISMS IN THE HUMAN GENOME

Since DNA polymorphisms promise to be so useful, it is obviously important to know at what frequency such variants occur in the genome, in other words to answer the question: how polymorphic is the human genome? Previous attempts to answer this question have relied upon electrophoretic and serological methods of detection of protein variants (Vogel & Motulsky, 1982). Using these now classical techniques, variability at the DNA level could only be inferred and estimated from protein data, for only the variability exhibited by coding sequences was measured. Moreover, only a small proportion of structural gene and related products was accessible to genetic analysis. Using the techniques to be described here, it is now possible not only to examine DNA polymorphisms directly but also to locate them in both the coding and non-coding portions of the genome.

The past few years have seen the isolation of a wide range of gene sequences and DNA segments from the human genome (Schmidtke & Cooper, 1983, 1984). Many of these DNA sequences have been shown to vary polymorphically (Cooper & Schmidtke, 1984). However, until recently, the only systematic study on the extent of DNA polymorphism was that of Jeffreys (1979) but this was restricted to a single locus, the β-globin gene cluster. Since the level of heterozygosity can be expected to differ between coding and non-coding sequences, Jeffreys' data are not necessarily representative of the genome as a whole. Furthermore, heterozygosity may also be expected to differ between different coding sequences depending upon the extent of conservation of those sequences. We have attempted to circumvent these problems by using DNA segments, cloned at random with respect to their coding potential, as probes to examine polymorphic variation in the human genome. Nineteen 'unique' cloned DNA segments, derived from flow-sorted metaphase chromosomes or total genomic DNA, were used as hybridisation probes in an analysis which screened nearly 27,000 base pairs of DNA sequence (Cooper & Schmidtke, 1984; Cooper et al, 1985). In all, fifteen different RFLPs were detected using these probes which were localized to chromosomes 7, 15, 21, 22 and X (Table 1). For every probe used, 10-15 individuals were screened with up to six different restriction enzymes. Figure 1 illustrates the principle involved. The presence of additional restriction enzyme sites (A' and B') around the region detected by the cloned probe alters the electrophoretic band patterns generated on a 'Southern blot'.

Variation was quantified using the single measure of heterozygosity,

$$h = 1 - \Sigma x_i^2$$

where x_i is the frequency of the ith allele. Each individual base-pair was considered to be one allele and the minimum number of base-pairs was

Table 1: Restriction fragment length polymorphisms detected by cloned single-copy genomic DNA segments

DNA Segment	Chromosome	Regional localization	Minimum number of base pairs screened with all enzymes	Restriction fragment length polymorphisms	Variant alleles	Allele frequencies of bands	Allele sizes (kilobase pairs)	Heterozygosity (h)
pJ 3.11	7	-	1636	Msp I Taq I	9 1	0.73/0.27 0.97/0.03	1.8/4.2 6.0/3.2	0.0122
pJ 5.11	7	-	2052	Msp I	6	0.70/0.30	5.5/10.0	0.0058
pB 78	7	-	1160	-	0	-	-	0.0000
pB 74	7+?	-	2496	Msp I Hind III Hind III Eco RI Bam HI	6 1 1 2 1	0.61/0.39 0.98/0.02 0.98/0.02 0.90/0.10 0.96/0.04	- /9.4 - /7.1 - /3.7 - /3.9 11.3/13.0	0.0088
pB 79a	7	-	1884	-	0	-	-	0.0000
pJU 48	7	-	1472	Taq I	1	0.97/0.03	8.2/15.6	0.0014
pJU 28	7	-	1488	-	0	-	-	0.0000
pJU 201	15	-	1628	Eco RI	11	0.58/0.42	1.9/1.8	0.0134
pAM 37	21	-	1456	Msp I	1	0.94/0.06	16.1/8.6	0.0014
λ 22.1	22	pter-q11	1680	Bam HI	3	0.90/0.10	13.5/20.5	0.0035
λ 22.3	22	q11-qter	1092	Bam HI Eco RI	2 2	0.92/0.08 0.92/0.08	22.9/20.0 7.3/7.1	0.0073
λ 22.4	22	q112-qter	480	-	0	-	-	0.0000
λ 22.6	22	q11-qter	1536	-	0	-	-	0.0000
pH 72	X	q28-qter	1764	-	0	-	-	0.0000
pJU 78	X	q24-q26	1098	-	0	-	-	0.0000
pB 20	X	q24-q26	1260	Msp I	2	0.95/0.05	5.2/ -	0.0032
pB 22	X	p11-q12	1168	-	0	-	-	0.0000
pB 97	Autosome	-	1236	-	0	-	-	0.0000
pX 2.7	X, 7	Xq24-qter	360	-	0	-	-	0.0000
			26946		49			

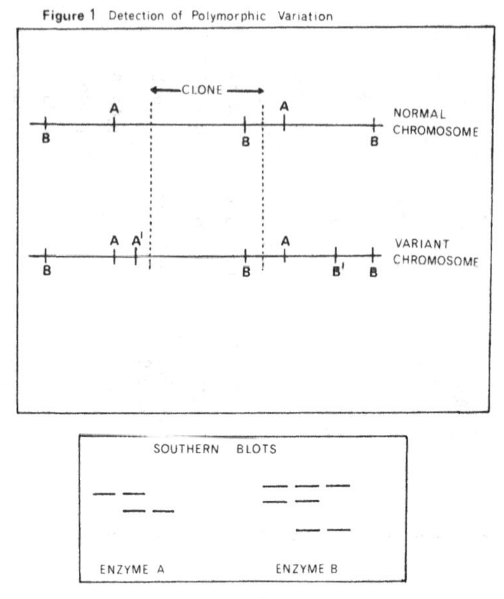

Figure 1: Detection of polymorphic variation

calculated as the product of the number of electrophoretic bands, the number of base-pairs in the restriction enzyme recognition sequence, and the number of chromosomes scored.

Average heterozygosity (H), calculated as the arithmetic mean of individual heterozygosities (Table 1) over all loci examined, was 0.0029. Calculation of the heterozygosity, (h), over all base-pairs examined gave the slightly higher figure of 0.0036. Average heterozygosity on autosomes was 0.0035 and on the X chromosome 0.0006. The total (0.0036) implies that about one in every 300 base-pairs in the human genome varies polymorphically. Since this estimate is an order of magnitude higher than Nei's (1975) estimate (0.0004) derived from protein data, most polymorphic variation present in the human genome must reside in non-coding sequences. Various *caveats* must, however, be mentioned. Heterozygosity would have been overestimated had probes detected several non-contiguous sequences in the genome because a greater number of restriction sites would have been screened. Conversely, heterozygosity would have been underestimated either when the fragment length variation lay below the limit of resolution or when small restriction fragments went undetected due to their inefficient binding to the nitrocellulose membrane. Possible limitations of our analysis include small sample size, both of the number of DNA regions examined and the number of base pairs within each region. It is not however thought likely that the results reported are affected greatly by these possible sources of error since the estimates of h are in reasonably good agreement with the heterozygosities calculated for the human β-globin (Jeffreys, 1979), serum albumin (Murray, et al, 1983) and parathyroid hormone (Schmidtke et al, 1984) loci. Obviously, the necessary selection of single or low copy sequences from the genome as hybridisation probes has meant that only unique DNA sequence heterozygosity was measured. Our estimate for heterozygosity in the human genome therefore needs to be substantiated by further work. Nevertheless, this study has placed certain limits upon the level of variation to be expected. In addition it provides an average figure with which to compare heterozygosities of other DNA sequences in the genome as these data begin to emerge.

MEDICAL APPLICATIONS

The use of polymorphisms as markers of genetic disease is not a novel idea. However, the classical approach of protein analysis was constrained by the necessity of utilizing the phenotypic effects of genetic variation. The number of detectable protein variants was low and many of them were not amenable to genetic analysis since they were not present in easily accessible tissues such as amniotic fluid cells or blood. The advent of recombinant DNA technology has promised to circumvent these problems since direct investigation of the genetic material obviates the requirement for specific tissue samples for analysis. Clinical diagnosis should now become much easier since a uniform methodology can be employed and clinical tests, once established, promise to be relatively easy, rapid and cheap. Direct analysis of the genetic defect using cloned gene probes is the most reliable and hence most desirable means of detection, and is clinically the most informative. However this has as yet only been applied to a relatively small number of genetic diseases. Alternative approaches to disease diagnosis involve the use of RFLPs, either associated with gene regions or with DNA segments linked to disease loci. The latter method requires no prior knowledge either of the primary gene product or of the basic biochemical mechanism of the disease.

For direct detection of the base-pair change or deletion event responsible for the disease phenotype, two conditions must be met. The nature of the disease must first be sufficiently well understood for the locus responsible to be identified e.g. globin gene defects in the thalassaemias. The next step is to isolate this locus by molecular cloning and to use the cloned gene as a

Table 2: Analysis of human genetic disease by means of recombinant DNA technology

Disease	Gene probe	Reference
Direct analysis of a genetic disease using gene probes to detect intragenic defects		
Acute lymphoblastic leukaemia	T-cell antigen receptor	Minden et al, 1985
Adrenal hyperplasia	Steroid 21-hydroxylase	White et al, 1984, 1985
Amyloidotic polyneuropathy	Albumin	Sasaki et al, 1985
Antithrombin III deficiency	Antithrombin III	Prochownicket al, 1983b
α_1-Antitrypsin deficiency	Synthetic oligonucleotide	Kidd et al, 1983
Atherosclerosis	Apolipoprotein A-1	Karathanasis et al, 1983b
Diabetes	Insulin	Haneda et al, 1983
Ehlers-Danlos syndrome	α1(1) Collagen	Pope et al, 1983
Glial tumours (?)	Epidermal growth factor	Libermann et al, 1985
Growth hormone deficiency	Growth hormone	Phillips et al, 1981
Haemophilia B	Factor IX	Giannelli et al, 1983
Hereditary persistance of foetal haemoglobin	β-Globin	Farquhar et al, 1983
Hypercholesterolaemia	Low density lipoprotein receptor	Lehrman et al, 1985
Hypoxanthine-guanine phosphoribosyltransferase	HPRT	Wilson et al, 1983
Lesch-Nyhan syndrome	HPRT	Yang et al, 1984; Nussbaum et al, 1983
Ornithine transcarbamylase deficiency	Ornithine trans-carbamylase	Rozen et al, 1985
Osteogenesis imperfecta	Pro α1(1) collagen	Chu et al, 1983; Pope et al, 1983
Retinoblastoma	Chromosome 13 DNA segments	Cavenee et al, 1983
Sickle cell anaemia	β-Globin Synthetic oligonucleotide	Geever et al, 1981; Conner et al, 1983
Tay Sach's disease	β-Hexosaminidase	Myerowitz & Proia, 1984
Thalassaemia	α- and β-globin	Orkin et al, 1978; Little et al, 1980
Wilm's tumour	Chromosome 11 DNA segments	Koufos et al, 1984

(?) denotes tentative report of rearrangement of EGF gene.

hybridisation probe to examine the allele(s) responsible for the disease phenotype. Mutations to be expected include single base-pair alterations and deletions which may interrupt, alter or otherwise interfere with any stage in the expression pathway. Table 2 illustrates the extent to which, though still in its infancy, the application of direct analytical techniques to disease diagnosis has already been applied. Since over two hundred different gene sequences have already been isolated from the human genome (Schmidtke & Cooper, 1983, 1984), a veritable armoury of gene probes is at present available which are potentially useful in the direct analysis of disease loci.

If a gene defect is not a gross deletion however, then it may often go undetected due to the lack of the suitable restriction enzyme (see Table 3). One alternative is to use DNA polymorphisms flanking the locus of interest as genetic markers. Inheritance of the disease allele can thus be monitored over the generations by following the inheritance of readily detectable RFLPs linked to the gene in question. Association of a specific allele with a genetic disease may also be excluded by analysis of the inheritance of linked RFLPs. The use of closely linked polymorphisms therefore permits diagnosis of specific genetic diseases in the absence of any causative explanation of the disease state. This strategy has now been used for the diseases listed in Table 4. As with direct detection of genetic lesions, this approach requires the possession of the appropriate gene probe. Yet for the majority of genetic diseases, the gene or genes responsible remain unknown. Diagnosis of such diseases requires a slightly different approach. Since the location of the mutant gene is unknown, direct analysis is impossible. Instead, linkage between a cloned DNA segment and the locus of interest is established and the inheritance of RFLPs associated with the linked DNA segment is investigated. The tighter the linkage between the DNA segment and the locus of interest, the smaller will be the number of recombinants as both loci tend to segregate together. However, since the phase of the markers is often unknown and disease analysis may be complicated by unknown variables such as incomplete or variable penetrance, likelihood analysis must be employed (Bishop & Skolnick, 1983). Briefly, a value of Θ, the recombination fraction, is derived such that the likelihood of obtaining the observed phenotypes is maximised. Comparison of the likelihood of linkage at given Θ with the likelihood of obtaining the observed phenotypes when the loci are unlinked ($\Theta = 0.5$) permits either confirmation or exclusion of linkage on a statistical basis (the "LOD score"). A list of genetic diseases for which linkage with polymorphic DNA segments has been established is given in Table 5. Unsuccessful attempts

Table 3: Studies which failed to detect gene defects using cloned gene probes

Disease	Probe	Reference
Adenine phosphoribosyl transferase (APRT) deficiency	APRT	Stambrook et al, 1984
Apolipoprotein CII deficiency	Apolipoprotein CII	Humphries et al, 1984; Fojo et al, 1984
Citrullinaemia*	Argininosuccinate synthetase	Su et al, 1962
Complement C2 deficiency	Complement C2	Cole et al, 1984
Congenital afibrinogenaemia	Fibrinogen	Uzan et al, 1984
Cystic fibrosis*	Complement C3	Davies et al, 1983a
Familial dysautonomia*	β-nerve growth factor	Breakefield et al, 1984
Hereditary fructose intolerance	Aldolase B	Grégori et al, 1984
Hypercholesterolaemia*	Apolipoprotein CII	Donald et al, 1985

* Linkage between DNA probe and disease locus excluded by analysis of inheritance of linked RFLPs.

have also been made to link specific polymorphisms to thyroxine-binding globulin deficiency (Hill, et al, 1982) and colour blindness (Murray et al, 1982). Most of the disease loci listed in Table 5 were already known to be situated on the X chromosome. The most notable achievement to date has been the localisation of the Huntington's disease gene to chromosome 4 (Gusella et al, 1983, 1984). The lack of polymorphic protein markers had previously prevented chromosomal localisation. Assuming that allelic heterogeneity for this disease can be excluded, additional DNA markers may be isolated for use as probes to determine whether individuals at risk for the disease are actually gene carriers.

Most of the DNA polymorphisms identified to date are due to the presence or absence of particular restriction sites, while deletions, insertions and copy number variation seem to occur less frequently (Cooper and Schmidtke, 1984). While RFLPs seem to be abundant in the genome, most exhibit a low allele frequency (Cooper and Schmidtke, 1984) and the task of identification of clinically useful variation remains quite laborious. For both gene mapping and gene linkage studies, it is necessary to possess a large number of polymorphic loci to identify each linkage group or chromosome. Although 150 polymorphic

Table 4: Indirect analysis of genetic disease using gene probes to detect closely linked polymorphisms.

Disease	Gene probes	Reference
α-Antitrypsin deficiency	α_1-Antitrypsin	Cox & Mansfield, 1985
Apolipoprotein CII deficiency	Apolipoprotein CII	Humphries et al, 1984
Atherosclerosis	Apolipoprotein A-1	Karathanasis et al, 1983a
Diabetes (Type II)	Insulin	Rotwein et al, 1981, 1983*
Growth hormone deficiency Type I	Growth hormone	Phillips et al, 1982
Haemophilia B	Factor IX	Giannelli et al, 1984; Grunebaum et al, 1984; Winship et al, 1984
Hypothyroidism	Thyroglobulin	Baas et al, 1984
Hypertriglyceridaemia	Apolipoprotein A-1	Rees et al, 1983
Sickle cell anaemia	β-Globin	Kan & Dozy, 1978; Phillips et al, 1980; Boehm et al, 1983
Lesch-Nyhan syndrome	HPRT	Nussbaum et al, 1983; Yang et al, 1983
Ornithine transcarbamylase deficiency	Ornithine transcarbamylase	Rozen et al, 1985; Old et al, 1985
Osteogenesis imperfecta	Pro α2(1) collagen	Tsipouras et al, 1984
Phenylketonuria	Phenylalanine hydroxylase	Woo et al, 1983
Thalassaemia	β-Globin	Boehm et al, 1983

* A firm association between diabetes and polymorphisms 5' to the insulin gene has been challenged by Yokoyama (1983).

DNA segments would be sufficient for this purpose if they were evenly spaced (Botstein et al, 1980), in practice some 1600 randomly selected segments will probably be required to ensure that a disease gene is certain to be linked to an RFLP locus at a distance no greater than 20 cM. Ideally, RFLPs should be frequent in the population under study in order to render the analysis applicable to a majority of patients. The existence of more than one RFLP at the disease locus is also advantageous. Each allele should segregate in true Mendelian fashion and the allelic distribution should follow Hardy-Weinberg expectations of equilibrium in the population. For indirect analysis, accuracy of diagnosis requires that the linkage between the marker sequence and the disease locus is

Table 5: Indirect analysis of genetic disease using cloned DNA segments to detect linked DNA polymorphisms

Disease	Probe	Distance between probe and disease locus (cM)	Reference
Fragile X - mental retardation syndrome	Factor IX	~30	Camerino et al, 1983; Choo et al, 1984; Warren et al, 1985 Zoll et al, 1985
Haemophilia A	DX13	<12	Harper et al, 1984
Huntington's chorea	G8	<10	Gusella et al, 1983
Menkes kinky hair	LI.28	16	Wieacker et al, 1983b
Muscular dystrophy			
Becker	LI.28	16	Kingston et al, 1983
Duchenne	λRC8	17	Davies et al, 1983a
	LI.28	17	Davies et al, 1983a
	Ornithine transcarbamylase	10	Davies et al, 1985
Myotonic dystrophy	Complement C3 gene	7	Davies et al, 1983b
Retinitis pigmentosa	LI.28	<15	Bhattacharya et al, 1984
Retinoschisis	λRC8	15	Wieacker et al, 1983c
Steroid-sulphatase-X-linked ichthyosis	λRC8	25	Wieacker et al, 1983a

very tight. Where linkage is loose, further isolation of flanking polymorphic DNA fragments on the other side of the locus of interest improves the accuracy of application. Clearly much effort needs to be spent identifying clinically useful RFLPs before such analysis becomes a routine procedure in medical genetics.

The initial localisation to a specific chromosome is sometimes regarded as the most difficult step and once this is achieved, the problem reduces to merely 'walking' along the genome to the locus of interest. Unfortunately, this is not so simple. Chromosome 'walking' refers to the isolation of overlapping fragments from a genomic library which can then in turn be used to isolate fragments still further away from the site of the original DNA probe. This process proceeds in steps of some 25 kb provided that no repetitive sequences are encountered. However, most linked DNA segments utilized to date are more than 10 cM from the disease locus (Table 5), a genetic distance that represents on average about 10^7 base-pairs. In addition, with increasing proximity to the desired locus, the detection of recombinants becomes more difficult so that the scale of the pedigree analysis required must increase exponentially. Clearly another approach is required.

Collins and Weissman (1984) have described an ingenious and potentially very important technique which should make it possible to 'jump' distances of up to 2000 kb in the direction of the disease locus. The basic principle is to generate very large genomic restriction fragments and to circularise these with DNA ligase. Smaller restriction fragments encompassing the marker locus, and including covalently linked DNA sequence originally located some considerable distance away, are then cloned and analysed. Once the location of the disease locus has been narrowed down, identification of the locus itself may be achieved by screening sequences isolated from this region for differential hybridisation to DNA from normal and affected individuals (e.g. Williamson et al, 1983) in the hope of finding a detectable deletion. It is too early yet to estimate the probability of success using this strategy.

Direct or indirect analysis as described above is also applicable to multifactorial genetic disease. Associations of particular polymorphic restriction fragments with disease states have already been observed for choriocarcinoma (Hoshina et al, 1984), insulin-dependent diabetes and multiple sclerosis (Cohen et al, 1984), although the relevance of such associations to a possible multifactorial etiology for these diseases remains unclear. Polymorphic DNA segments are also useful in helping to define the deletions and translocations that can play a part in oncogenesis (Naylor et al, 1984; Ellis & Davies, 1985). Furthermore, segments have recently been used to determine the host or donor origin of cell populations after bone marrow transplantation (Ginsberg et al, 1985). Other uses for DNA polymorphisms include paternity testing and population studies (e.g. Roberts, 1982).

Using recombinant DNA methodology, the abundant polymorphic variation present in human populations can be tapped for gene mapping and linkage studies and utilised to provide greater insight into pathological processes leading to a better understanding of the molecular basis of inherited disease. Clearly, we have only just begun to exploit the potential of DNA polymorphisms in clinical medicine.

OTHER USES

Clinical applications have been stressed, since it is here that the practical use of RFL polymorphisms is likely to be most rewarding. However, with over

100 high frequency RFLPs in human genomic DNA (autosomal and gonosomal) known today and with the corresponding cloned DNA sequences available, there are numerous other applications. The first task will be to assess the frequency of the RFL polymorphisms in different populations. In particular, the small populations under threat of extinction who have long been isolated should come in for urgent attention. Once these frequencies are established, they can be used in the same way as classical markers for studies of genetic affinity, genetic distance, kinship, paternity, zygosity and similarity. Indeed, all the purposes for which classical markers have been used can be examined. One such has been given above - the extent of heterozygosity. There is no reason, moreover, to regard the linkage of RFL polymorphisms over all populations to be constant, and there may be differences in linkage disequilibria from one population to another. This will be of direct relevance in the utilisation of linkages of the RFLPs to disease loci. Associations of these polymorphisms with disease, their role in susceptibility to disease, and hence their selective advantages and disadvantages deserve scrutiny. Chromosome evolution, using chromosomally and regionally allocated clone sequences, can be expected to throw light on the phylogeny of human populations and, on a broader canvas, of mankind and the higher primates generally.

The work so far has concentrated essentially on subjects of western European, Japanese, and American Black affinity. From what is known of classical gene markers, these represent but a fraction of the total genetic heterogeneity that exists in man. To that heterogeneity, tropical populations make a predominant contribution. It will be interesting to see whether the same applies in RFL polymorphism studies.

ACKNOWLEDGMENTS

We wish to thank Louis Lim and Christine Hall for criticising the manuscript and Daksha Gandhi for typing it. The support of the Wellcome Trust (D.N.C.) and the Deutsche Forschungsgemeinschaft (J.S.) is gratefully acknowledged.

REFERENCES

Baas, F., Bikker, H., Van Ommen, G. & Vijlder, J. (1984). Unusual scarcity of restriction site polymorphism in the human thyroglobulin gene. A linkage study suggesting autosomal dominance of a detective thyroglobulin allele. Human Genetics, **67**, 301-305.

Bhattacharya, S. S., Wright, A. F., Clayton, J. F., Price W. H., Philips, C. I., McKeown, C.M.E., Jay, M., Bird, A. C., Pearson, P. L., Souther, E. M. & Evans, H. J. (1984). Close genetic linkage between X-linked retinitis pigmentosa and a restriction fragment length polymorphism identified by recombinant DNA probe L1.28. Nature, **309**, 253-255.

Bishop, D. T. & Skolnick, M. H. (1983). Genetic markers and linkage analysis. In: Banbury Report, 14, Recombinant DNA Applications to Human Disease, pp. 251-259. Cold Spring Harbor Laboratory.

Boehm, C. D., Antonarakis, S. E., Phillips, J. A., Stetten, G. & Kazazian, H. H. (1983). Prenatal diagnosis using DNA polymorphisms. Report on 95 pregnancies at risk for sickle-cell disease or β-thalassaemia. New England Journal of Medicine, **308**, 1054-1058.

Botstein, D., White, R. L., Skolnick, M. & Davis, R. W. (1980). Construction of a genetic linkage map in man using restriction fragment length polymorphisms. American Journal of Human Genetics, **32**, 314-331.

Breakefield, X. O., Orloff, G., Castoglione, C., Coussens, L., Axelrod, F.B. & Ullrich, A. (1984). Structural gene for β-nerve growth factor not defective in familial dysautonomia. Proceedings of the National Academy of Sciences, U.S.A., **81**, 4213-4216.

Camerino, G., Mattei, M. G., Mattei, J. F., Jaye, M. & Mandel, J. L. (1983). Close linkage of fragile X-mental retardation syndrome to Haemophilia B and transmission through a normal male. Nature, **305**, 779-784.

Choo, K. G., George, D., Filby, G., Halliday, J. L., Leversha, M., Webb, G. & Danks, D. M. (1984). Linkage analysis of X-linked mental retardation with and without fragile-X using factor IX gene probe. Lancet, ii, 349.

Chu, M. L., Williams, C. J., Pepe, G., Hirsch, J. L., Prockop, D. J. & Ramirez, F. (1983). Internal deletion in a collagen gene in a perinatal lethal form of osteogenesis imperfecta. Nature, **304**, 78-80.

Cohen, D., Cohen, O., Marcadet, A., Massart, C., Lathrop, M., Deschamps, I., Hors, J., Schuller, E. & Dausset, J. (1984). Class II HLA-DC β-chain DNA restriction fragments differentiate among HLA-DR2 individuals in insulin-dependent diabetes and multiple sclerosis. Proceedings of the National Academy of Sciences, U.S.A., **81**, 1774-1778.

Cole, F. S., Whitehead, A. S., Auerbach, H. S., Perlmutter, D. H., Lint, T. F., Zeitz, H. J. & Colten, H. R. (1984). Molecular basis of the genetic deficiency of the second component of complement (C2) in the human. Journal of Cell Biology, **99**, 330a.

Collins, F. S. & Weissman, S. M. (1984). Directional cloning of DNA fragments at a large distance from an initial probe: a circularization method. Proceedings of the National Academy of Sciences, U.S.A., **81**, 6812-6816.

Conner, B. J., Reyes, A. A., Morin, C., Itakura, K., Teplitz, R. L. and Wallace, R. B. (1983). Detection of sickle cell β^S-globin allele by hybridization with synthetic oligonucleotides. Proceedings of the National Academy of Sciences, U.S.A., **80**, 278-282.

Cooper, D. N. & Schmidtke, J. (1984). DNA restriction fragment length polymorphisms and heterozygosity in the human genome. Human Genetics, **66**, 1-16.

Cooper, D. N., Smith, B. A., Cooke, H. J., Neimann, S. and Schmidtke, J. (1985). An estimate of unique DNA sequence heterozygosity in the human genome. Human Genetics, **69**, 201-205.

Cox, D. W. & Mansfield, T. (1985). Prenatal diagnosis for alpha 1-antitrypsin deficiency. Lancet, **i**, 230.

Davies, K. E., Pearson, P. L., Harper, P. S., Murray, J. M., O'Brien, T. O., Sarfarazi, M. & Williamson, R. (1983a). Linkage analysis of two cloned DNA sequences flanking the Duchenne muscular dystrophy locus on the short arm of the human X chromosome. Nucleic Acids Research, **11**, 2303-2312.

Davies, K. E., Jackson, J., Williamson, R., Harper, P. S., Ball, S., Sarfarazi, M., Meredith, L. & Fey, G. (1983b). Linkage analysis of myotonic dystrophy and sequences on chromosome 19 using a cloned complement 3 gene probe. Journal of Medical Genetics, **20**, 259-263.

Davies, K.E., Gillian, T.C. and Williamson, R. (1983c). Cystic fibrosis is not caused by a defect in the gene coding for human complement C3. Molecular Biology and Medicine, **1**, 185-190.

Davies, K.E., Briand, P., Ionasescu, V., Ionasescu, G., Williamson, R., Brown, C., Cavard, C. & Cathelineau, L. (1985). Gene for OTC: characterization and linkage to Duchenne muscular dystrophy. Nucleic Acids Research, **13**, 155-165.

Donald, J. A., Wallis, S. C., Kessling, A., Tippett, P., Robson, E. B., Ball, S., Davies, K. E., Scambler, P., Berg, K., Heiberg, A., Williamson, R. & Humphries, S. E. (1985). Linkage relationships of the gene for apolipoprotein CII with loci on chromosome 19. Human Genetics, **69**, 39-43.

Ellis, K. P. & Davies, K. E. (1985). An appraisal of the application of recombinant DNA techniques to chromosome defects. Biochemical Journal, **226**, 1-11.

Farquhar, M., Gelinas, R., Tatsis, B., Murray, J., Yagi, M., Mueller, R. & Stamatoyannopoulos, G. (1983). Restriction endonuclease mapping of γ-σ-β-globin region in $^{G}\gamma(\beta)^{+}$ HPFH and a Chinese $^{A}\gamma$HPFG variant. American Journal of Human Genetics, **35**, 611-620.

Fojo, S. S., Law, S. W., Sprecher, D. L., Gregg, R. E., Baggio, G. & Brewer, H. B. (1984). Analysis of the Apo C-II gene in Apo C-II deficient patients. Biochemical and Biophysical Research Communications, **124**, 308-313.

Geever, R. F., Wilson, L. B., Nallaseth, F. S., Milner, P. F., Bittner, M. & Wilson, J. T. (1981). Direct identification of sickle cell anemia by blot hybridization. Proceedings of the National Academy of Sciences, U.S.A., **78**, 5081-5085.

Giannelli, F., Choo, K. H., Rees, D. J. G., Boyd, Y., Rizza, C. R. & Brownlee, G. G. (1983). Gene deletions in patients with haemophilia B and anti-factor IX antibodies. Nature, **303**, 181-182.

Giannelli, F., Choo, K. H., Winship, P. R., Rizza, C. R., Anson, D. S., Rees, D. J. G., Ferrari, N. & Brownlee, G. G. (1984). Characterization and use of an intragenic polymorphic marker for detection of carriers of haemophilia B (factor IX deficiency). Lancet, **i**, 239-249.

Ginsburg, D., Antin, J. H., Smith, B. R., Orkin, S. H. & Rappeport, J. M. (1985). Origin of cell populations after bone marrow transplantation. Analysis using DNA sequence polymorphisms. Journal of Clinical Investigation, **75**, 596-602.

Grégori, C., Besmond, C., Odievre, M., Kahn, A. & Dreyfus, J. C. (1984) DNA analysis in patients with hereditary fructose intolerance. Annals of Human Genetics, **48**, 291-296.

Grunebaum, L., Cazenave, J. -P., Camerino, G., Kloepfer, C. & Mandel, J. -L. (1984). Carrier detection of Haemophilia B by using a restriction site polymorphism associated with the coagulation factor. Journal of Clinical Investigation, **73**, 1491-1495.

Gusella, J. F., Wexler, N. S., Conneally, P. M., Naylor, S. L., Anderson, M. A., Tanzi, R. E., Watkins, P. C., Ottina, K., Wallace, M. R., Sakaguchi, A. Y., Young, A. B., Shoulson, I., Bonilla, E. & Martin, J. B. (1983). A polymorphic DNA marker genetically linked to Huntington's disease. Nature, **306**, 234-238.

Gusella, J. F., Tanzi, R. E., Anderson, M. A., Hobbs, W., Gibbons, K., Raschtchian, R., Gillian, T. C., Wallace, M. R., Wexter, N. W. & Conneally, P. M. (1984). DNA markers for nervous system diseases. Science, **225**, 1320-1326.

Haneda, M., Chan, S. J., Kwok, S. C. M., Rubenstein, A. H. & Steiner, D. F. (1983). Studies on mutant human insulin genes: identification and sequence analysis of a gene encoding (Ser^{B24}) insulin. Proceedings of the National Academy of Sciences, U.S.A., **80**, 6366-6370.

Harper, K., Pembrey, M. E., Davies, K. E., Winter, R. M., Hartley, D. & Tuddenham, E. G. D. (1984). A clinically useful DNA probe closely linked to Haemophilia A. Lancet, ii, 6-8.

Hill, M. E. E., Davies, K. E., Harper, P. & Williamson, R. (1982). The Mendelian inheritance of a human X chromosome-specific DNA sequence polymorphism and its use in linkage studies of genetic disease. Human Genetics, **60**, 222-226.

Hoshina, M., Boothby, M.R., Hussa, R.D., Pattillo, R. A., Camel, H. M. and Boime, I. (1984). Segregation patterns of polymorphic restriction sites of the gene encoding the α-subunit of human chorionic gonadotrophin in trophoblastic disease. Proceedings of the National Academy of Sciences, U.S.A., **81**, 2504-2507.

Humphries, S. E., Williams, L., Myklebost, D., Stalenhoef, A. F. H., Demacker, P. N. M., Baggio, G., Crepaldi, G., Galton, D.J. & Williamson, R. (1984). Familial apolipoprotein CII deficiency: a preliminary analysis of the gene defect in two independnet families. Human Genetics, **67**, 151-155.

Jeffreys, A. J. (1979). DNA sequence variants in the $^{G}\gamma$-, $^{A}\gamma$, δ and β-globin genes of man. Cell, **18**, 1-10.

Kan, Y. W. & Dozy, A. M. (1978). Polymorphisms of DNA sequence adjacent to human β-globin structural gene: relationship to sickle mutation. Proceedings of the National Academy of Sciences, U.S.A., **75**, 5631-5635.

Karathanasis, S. K., Norum, R. A., Zannis, V. I. & Breslow, J. L. (1983a). An inherited polymorphism in the human anpolipoprotein A-1 gene locus related to the development of atherosclerosis. Nature, **301**, 718-720.

Karanthanasis, S. K., Zannis, V. I. & Breslow, J. L. (1983b). A DNA insertion in the apolipoprotein A-I gene of patients with premature atherosclerosis. Nature, **305**, 823-825.

Kidd, V. J., Wallace, R. B., Itakura, K. & Woo, S. L. C. (1983). α_1 antitrypsin deficiency detection by direct analysis of the mutation in the gene. Nature, **304**, 230-234.

Koufos, A., Hansen, M. F., Lampkin, B. C., Workman, M. L., Copeland, N. G., Jenkins, N. A. & Cavenee, W. K. (1984). Loss of alleles at loci on human chromosome 11 during genesis of Wilm's tumour. Nature, **309**. 170-172.

Kingston, H. M., Thomas, N. S. T., Pearson, P. L., Sarfarazi, M. & Harper, P. S. (1983). Genetic linkage between Becker muscular dystrophy and a polymorphic DNA sequence on the short arm of the X chromosome. Journal of Medical Genetics, **20**, 255-258.

Lehrman, M. A., Schmeider, W. J., Sudhof, T. C., Brown, M. S., Goldstein, J. L. & Russell, D. W. (1985). Mutation in LDL receptor: Alu-Alu recombination deletes exons encoding transmembrane and cytoplasmic domains. Science, **227**, 140-146.

Libermann, T. A., Nussbaum, H. R., Razon, N., Kris, R., Lax, I., Soreq, H., Whittle, N., Waterfield, M. D., Ullrich, A. & Schlessinger, J. (1985). Amplification, enhanced expression and possible rearrangement of EGF receptor gene in primary human brain tumours of glial origin. Nature, **313**, 144-147.

Little, P. F. R., Annison, G., Darling, S., Williamson, R., Camba, L. & Modell, B. (1980). Model for antenatal diagnosis of β-thalassaemia and other monogenic disorders by molecular analysis of linked DNA polymorphisms. Nature, **285**, 144-147.

Minden, M. D., Toyonaga, B., Ha, K., Yanagi, Y., Chin, B., Gelfand, E. & Mak, T. (1985). Somatic rearrangement of T-cell receptor gene in human T-cell malignancies. Proceedings of the National Academy of Sciences, U.S.A., **82**, 1224-1227.

Murray, J. M. Davies, K. E., Harler, P. S., Meredith, L., Mueller, C. R. & Williamson, R. (1982). Linkage relationship of a cloned DNA sequence on the short arm of the X chromosome to Duchenne muscular dystrophy. Nature, **300**, 69-71.

Murray, J. C., Demopulos, C. M., Lawn, R. M. & Motulsky, A. G. (1983). Molecular genetics of human serum albumin: restriction enzyme fragment length polymorphisms and analbuminemia. Proceedings of the National Academy of Sciences, U.S.A., **80**, 5951-5955.

Myerowitz, R. & Proia, R. L. (1984). cDNA clone for the α-chain of human β-hexosaminidase: deficiency of α-chain in Ashkenazi Tay-Sachs fibroblasts. Proceedings of the National Academy of Sciences, U.S.A., **81**, 5394-5398.

Naylor, S. L., Minna, J., Johnson, B. & Sakaguchi, A. Y. (1984). DNA polymorphisms confirm the deletion in the short arm of chromosome 3 in small cell lung cancer. American Journal of Human Genetics, **36**, 35S.

Nei, M. (1975). Molecular population genetics and evolution. Amsterdam: North Holland.

Nussbaum, R. L., Crowder, W. E., Nyhan, W. L. & Caskey, C. T. (1983). A three allele restriction-fragment-length polymorphism at the hypoxanthine phosphoribosyltransferase locus in man. Proceedings of the National Academy of Sciences, U.S.A., **80**, 4035-4039.

Old, J. M., Purvis-Smith, S., Wilcken, B., Pearson, P., Williamson, R., Briand, P. L., Howard, N. J., Hammond, J., Cathelineau, L. & Davies, K. E. (1985). Prenatal exclusion of ornithine transcarbamylase deficiency by direct gene analysis. Lancet, **i**, 73-75.

Orkin, S. H., Alter, B. P., Altay, C., Mahoney, M. J., Lazarus, H., Hobbins, J. C. & Nathan, D. G. (1978). Application of endonuclease mapping to the analysis and prenatal diagnosis of thalassemias caused by globin-gene deletion. New England Journal of Medicine, **299**, 166-172.

Phillips, J. A., Panny, S. R., Kazazian, H. H., Boehm, C. D., Scott, A. F. & Smith, K. D. (1980). Prenatal diagnosis of sickle cell anemia by restriction endonuclease analysis: Hind III polymorphisms in γ-globin genes extend test applicability. Proceedings of the National Academy of Sciences, U.S.A., **77**, 2853-2856.

Phillips, J. A., Hjelle, B. L., Seeburg, P. H. & Zachmann, M. (1981). Molecular basis for familial islated growth hormone deficiency. Proceedings of the National Academy of Sciences, U.S.A., **78**, 6372-6375.

Phillips, J. A., Parks, J. S., Hjelle, B. L., Herd, J. E., Plotnick, L. P., Migeon, C. J. & Seeburg, P. H. (1982). Genetic analysis of familial isolated growth hormone deficiency type 1. Journal of Clinical Inviestigation, **70**, 489-495.

Pope, F. M., Nicholls, A. C. & Grosveld, F. G. (1983). Similar αI(1)-like gene deletions cause some types of Ehlers Danlos syndrome type II and lethal osteogenesis imperfecta. Clinical Genetics, **24**, 303.

Prochownik, E. V., Antonarakis, S., Bauer, K. A., Rosenberg, R. D., Fearon, E. R. & Orkin, S. H. (1983). Molecular heterogeneity of inherited antithrombin III deficiency. New England Journal of Medicine, **308**, 1549-1552.

Rees, A., Shoulders, C. C., Stocks, J., Galton, D. J. & Baralle, F. E. (1983). DNA polymorphism adjacent to human apoprotein A-1 gene: relation to hypertriglyceridemia. Lancet, i, 444.

Roberts, D. F. (1982). Applications of polymorphisms in anthropogenetic studies. Human Biology, **54**, 175.

Rotwein, P., Chyn, R., Chirgwin, J., Cordell, B., Goodman, H. M. & Permutt, M. A. (1981). Polymorphism in the 5'-flanking region of the human insulin gene and its possible relation to type 2 diabetes. Science, **213**, 1117-1120.

Rotwein, P. S., Chirgwin, J., Province, M., Knowler, W. C., Pekitt, D. J., Cordell, B., Goodman, H. M. & Pettitt, M. A. (1983). Polymorphism in the 5' flanking region of the human insulin gene: a genetic marker for non-insulin-dependent diabetes. New England Journal of Medicine, **308**, 65-71.

Rozen, R., Fox, J., Fenton, W. A., Horwich, A. L. & Rosenberg, L. E. (1985). Gene deletion and restriction fragment length polymorphisms at the human ornithine carbamylase locus. Nature, **313**, 815-817.

Sasaki, H., Sakaki, Y., Takagi, Y., Sahashi, K., Takahashi, A., Isobe, T., Shinoda, T., Matsuo, H., Goto, I. & Kuroiwa, T. (1985). Presymptomatic diagnosis of heterozygosity for familial amyloidotic polyneuropathy by recombinant DNA techniques. Lancet, i, 100.

Schmidtke, J. & Cooper, D. N. (1983). A list of cloned human DNA sequences. Human Genetics, **65**, 19-26.

Schmidtke, J. & Cooper, D. N. (1984). A list of cloned human DNA sequences-supplement. Human Genetics, **67**, 111-114.

Schmidtke, J., Pape, B., Krengel, U., Langenbeck, U., Cooper, D. N., Breyel, E. & Mayer, H. (1984). Restriction fragment length polymorphisms at the human parathyroid hormone gene locus. Human Genetics, **67**, 428-431.

Stambrook, P. J., Dush, M. K., Trill, J. J. & Tischfield, J. A. (1984). Cloning of a functional human adenine phosphoribosyltransferase (APRT) gene:

identification of a restriction fragment length polymorphism and preliminary analysis of DNAs from APRT-deficient families and cell mutants. Somatic Cell Molecular Genetics, **10**, 359-367.

Tsipouras, P., Børresen, A.-L., Dickson, L.A., Berg, K., Prockop, D. J. & Ramirez, F., (1984). Molecular heterogeneity in the mild autosomal dominant forms of osteogenesis imperfecta. American Journal of Human Genetics, **36**, 1172-1179.

Uzan, G., Courtois, G., Besmond, C., Frain, M., Sala-Trepat, J., Kahn, A. & Marguerie, G. (1984). Analysis of fibrinogen genes in patients with congenital afibrinogenemia. Biochemical and Biophysical Research Communications, **120**, 376-383.

Vogel, F. & Motulsky, A. G. (1982). Human Genetics, 2nd edn. Berlin: Springer.

Warren, S. T., Glover, T. W., Davidson, R. L. & Jagadeeswaran. P. (1985). Linkage and recombination between fragile X-linked mental retardation and the factor IX gene. Human Genetics, **69**, 44-46.

White, P. C., New, M. I. & Dupont, B. (1984). HLA-linked congenital adrenal hyperplasia results from a defective gene encoding a cytochrome P-450 specific for steroid 21-hydroxylation. Proceedings of the National Academy of Sciences, U.S.A., **81**, 7505-7509.

White, P. C., Grossberger, D., Onufer, B. J., Chaplin, D. D., New, M. I., Dupont, B. & Strominger, J. L. (1985). Two genes encoding steroid 21-hydroxylase are located near the genes encoding the fourth component of complement in man. Proceedings of the National Academy of Sciences, U.S.A., **82**, 1089-1093.

White, R., Leppert, M., Bishop, D. T., Barker, D., Berkowitz, J., Brown, C., Callahan, P., Holm, T. & Jerominski, L. (1985). Construction of linkage maps with DNA markers for human chromosomes. Nature, **313**, 101-105.

Wieacker, P., Davies, K. E., Mevorah, B. & Ropers, H. H. (1983a). Linkage studies in a family with X-linked recessive ichthyosis employing a cloned DNA sequence from the distal short arm of the X chromosome. Human Genetics, **63**, 113-116.

Wieacker, P., Horn, N., Pearson, P., Wienker, T. F., McKay, E. and Ropers, H. H. (1983b). Menkes kinky hair disease: a search for closely linked restriction fragment length polymorphisms. Human Genetics, **64**, 139-142.

Wieacker, P., Wienker, T. F., Dallapiccola, B., Bender, K., Davies, K. E. & Ropers, H. H. (1983c). Linkage relationships between retinoschisis, Xg, and a cloned DNA sequence from the distal short arm of the X chromosome. Human Genetics, **64**, 143-145.

Williamson, R., Crampton, J. M., Clarke, B. E., Davies, K. E., Knapp, T. F., & Woods, D. (1983). The study of cystic fibrosis using gene probes. In: Banbury Report, 14, Recombinant DNA Applications to Human Disease, Cold Spring Harbor Laboratory.

Wilson, J. M., Young, A. B. & Keley, W. N. (1983). Hypoxanthine-guanine phosphoribosyltransferase deficiency: the molecular basis of the clinical syndromes. New England Journal of Medicine, **309**, 900-910.

Winship, P. R., Anson, D. S., Rizza, C. R. & Brownlee, G. G. (1984). Carrier detection in haemophilia B using two further intragenic RFLPs. Nucleic

Acids Research, **12**, 8861-8872.

Woo, S. L. C., Lidsky, A. S., Guttler, F., Chandra, T. & Robson, K. J. H. (1983). Cloned human phenylalanine hydroxylase gene allows prenatal diagnosis and carrier detection of classical phenylketonuria. Nature, **306**, 151-155.

Yang, T. P., Patel, P. I., Brennand, J., Chinault, A. C. & Laskey, C. T. (1983). Molecular analysis of the human HPRT locus. Anstracts 34th Annual Meeting, American Society of Human Genetics, 185A.

Yang, T. P., Patel, P. I., Chinault, A. C., Stout, J. T., Jackson, L. G., Hildebrand, B. M. & Caskey, C. T. (1984). Molecular evidence for new mutation at the HPRT locus in Lesch-Nyhan patients. Nature, **310**, 412-414.

Yokoyama, S. (1983). Polymorphism in the 5'-flanking region of the human insulin gene and the incidence of diabetes. American Journal of Human Genetics, **35**, 193-200.

Zoll, B., Arnemann, J., Cooper, D. N., Krawczak, M., Pescia, G., Wahli, W. & Schmidtke, J. (1985). Evidence against close linkage of the loci for fra Xq of Martin-Bell syndrome and factor IX. Human Genetics, **71**, 122-126.

Note added in proof (April 1986):

Data included in the tables contain reports up to March 1985. A more detailed review article by the same authors entitled "Diagnosis of genetic disease using recombinant DNA" is now in press in Human Genetics. The latest sequence list newsletter, Gene Communications Vol.2 no.1, contains some 2,500 independent reports of cloned human DNA sequences and is available from the authors.

NUCLEOTIDE SEQUENCES, RESTRICTION MAPS, AND HUMAN MITOCHONDRIAL DNA DIVERSITY

R. L. CANN

Howard Hughes Medical Institute, University of California, San Francisco, U.S.A.

INTRODUCTION

A great deal has been learned in the last five years about the molecular basis of human mitochondrial DNA polymorphisms (Brown, 1980; Giles et al, 1980; Cann & Wilson, 1983; Greenberg et al, 1983; Johnson et al, 1983; Horai et al, 1984; Cann, Brown & Wilson, 1984). This new field has contributed to issues of mutation rates, DNA repair, and ageing or programmed cell death, topics important in human evolution no matter which population is studied. However, mitochondrial DNA (mtDNA) also carries the potential for unifying information on maternal lineages into convenient markers for phenotypic and organism diversity, where one is concerned with sexual selection, founding events, migration, admixture, and cultural revolutions. Some tropical populations by virtue of their relative isolation or historical accidents in contact with western Europeans, provide unique opportunities to study the human gene pool and how it responds in periods of cultural transition, particularly the female contribution to it.

This paper summarises information derived from the study of human mitochondrial polymorphisms on a worldwide basis. Some of the polymorphisms are private, some are cosmopolitan. The classification of these may depend simply on the level at which one works, i.e. DNA sequencing, high resolution restriction fragment mapping, or Southern blots of genomic DNA using purified mtDNA as a molecular probe. Knowledge is at the stage where one can begin to tailor the technology to the question asked. This may be as global as enquiring how reproductive strategies differ between males and females, or may be as limited as asking how many canoes eventually landed in Samoa. Uniparental inheritance, as in mtDNA transmission through female lineages, allows these questions of evolution and adaptation to be investigated at diverse levels.

Table 1: Characteristics of mitochondrial DNA

Maternally inherited
Circular, 16.5 kilobases
High copy number (1000-5000/cell)
Contained within mitochondria
Homogeneous within an individual
Has rRNAs, tRNAs, protein-coding and non-coding regions
Does not appear to recombine
Limited to no-repair capacity
Different genetic code from nucleus
Evolves primarily by single base substitution

MITOCHONDRIAL GROUND RULES

Table 1 summarises some essential points about mtDNA, and other aspects are discussed elsewhere (Brown, 1983; Wilson et al, 1985; Chomyn et al, 1985). For present purposes, it is important to concentrate on the small size, the relative homogeneity within an individual but heterogeneity between individuals, and the rate of mutation. Although there are many copies of this DNA in each tissue or organ, all copies look like a clone, or nearly exact replicas of each other, with a few important exceptions. These exceptions provide avenues for testing hypotheses about mutation rates and repair of DNA damage.

While most of the genes for mitochondrial proteins are encoded by the nucleus, the mitochondrial genome has a small but diverse class of genes itself. As in the nucleus, there are ribosomal genes, transfer RNA genes, genes which code for proteins that take part in the production of energy within the mitochondria, and a few small noncoding regions which are important for DNA replication. Unlike genes in the nucleus, animal mitochondrial genes have some additional features. They are read with a slightly different genetic code. They do not appear to recombine with each other or with cryptic homologous sequences in the nucleus, and there does not appear to be an active DNA repair system. How this all relates to their being transmitted by females is not yet understood.

New mutations in the mitochondrial gene pool, which are now known to occur primarily by single base changes, give almost every one of us our own mitochondrial fingerprint. In order to use this property as a clock for examining human divergence and diversity, it is necessary to know how fast mtDNA evolves. As a crude estimate in primates, the rate at which the whole genome

Table 2: Properties of human mitochondrial DNA evolution deduced from restriction mapping studies

High level of polymorphism in human gene pool
Sequence evolution by both single nucleotide changes and small (1-3 bp) additions or deletions
Maternal inheritance
No strict association of polymorphism with race
Parallel evolution of many restriction site polymorphisms in different populations

evolves is about 2% per million years, or about 10 times the rate of the average single copy nuclear gene of the same functional class (Brown et al, 1979). This rate was measured in the hominoid apes at Berkeley, and is roughly linear for the first 10 million years of divergence (Brown et al, 1982). Human gene pools diverging some time after the origin of the genus *Homo* two million years ago show differences that are well within the linear range of evolution of the molecule. There are problems which occur if two species have diverged so much that the same sites are undergoing multiple substitutions, giving a plateau effect in the rate of change for mtDNA.

ANALYSIS OF HUMAN mtDNA BY RESTRICTION ENDONUCLEASES

Restriction enzymes were first used to probe human mitochondrial polymorphisms beginning in 1975. By 1982 the regions were known in which humans were free to vary in sequence, and these could be contrasted with regions where no substitutions were ever seen. The picture of human mtDNA variation was filling out, and examination of a very large number of sites, over 460, showed that these point mutations were quite evenly scattered. Restriction studies provide useful and necessary information, some of which is summarised in Table 2.

These studies alerted us to a high level of hidden polymorphism in the human gene pool, hidden in the sense that few mitochondrial proteins showed allelic variation in population surveys. The mitochondria were thought to be the domain of metabolically sensitive and therefore evolutionarily conservative proteins. Most of these polymorphisms are due to single base substitutions, and most are silent. They affect the third position in a codon and do not change the amino acid sequence of a protein, and so appear to be neutral mutations in some respects. Another interesting class was shown by high resolution mapping to be due to tiny length mutations, or small deletions and additions.

Limited family studies demonstrated maternal inheritance of restriction polymorphisms for mtDNA, and while a small number of these polymorphisms could be classified as private, or highly localised in their distribution, most were widely dispersed in the human population. In a composite gene map of the variation seen in just 12 Australian Aborigines from western Australia, a small sample of what must have been a much larger and more diverse population before European contact, many of the same mitochondrial mutations identified can be found in Rome, Beijing, Lagos or San Francisco (Cann, 1985).

In a phylogenetic analysis, where shared derived restriction site changes are used to construct a diagram of the human population and how it diverged, one can demonstrate that many of these polymorphisms must have evolved apparently independently in different parts of the world (Cann & Wilson, 1983; Cann, Stoneking & Wilson, unpublished). While it was thought that they represented unique events which would indicate close genetic connections, this would lead to a classification of genotypes. But if a restriction polymorphism could arise due to a site loss by mutation in any one of six bases within the recognition sequence of the enzyme then they would lead to a classification of molecular phenotypes, not necessarily genotypes; it is easy to see how one might generate errors in the reconstruction of human genealogies. Also, could males occasionally contribute their mitochondrial genes?

Sequencing studies

For these and other reasons, we developed further more intensive human mitochondrial polymorphism studies, and started to clone and base sequence the mtDNA of a few individuals, as did other laboratories (Greenberg et al, 1983; Oliver et al, 1983; Monat & Loeb, 1985). There are now many data on human mtDNA sequences, in addition to the one genome completely sequenced by Sanger's group at Cambridge (Anderson et al, 1981).

Fortunately for human geneticists, the rapid cloning and sequencing system with the M13 vector has taken much of the labour out of this intensive analysis (Messing, 1983). To date, almost 8 kilobases of sequence are available for comparison to mtDNA purified at Berkeley. I have concentrated on five individuals for the most part, two Australians, a Tongan, and two Blacks. The amount actually sequenced for each is listed in Table 3, and represents regions of all functional classes within the mitochondrial genome. The amount of sequence divergence, in percent compared to the one complete reference sequence, is

Table 3: Phylogenetic comparisons from mtDNA coding sequences

Individual	Source	Base-pairs sequenced	% divergence from Cambridge Reference Sequence*
American Black (Denver)	placenta	922	0.76
American Black (HeLa)**	tissue culture cells	896	0.22
Nigerian	placenta	157	0
Australian Aborigine (Perth)	placenta	1887	0.32
Australian Aborigine (Derby-Broome)	placenta	1934	0.21
South African Aboriginal	tissue culture cells	1139	0.61
Tongan	placenta	757	0.40
Finn	placenta	218	0.92
		Total 7910	Average 0.43

* Anderson et al., 1981.

** Brown et al., 1982.

listed for each. On average, these human mtDNA sequences are about 0.43% divergent.

What kind of variation was uncovered by this intense enquiry? One of the properties often remarked upon by physiologists who study Australians is their ability to tolerate abrupt environmental shifts in temperature. This is usually attributed to acclimatisation, but genes which are concerned with internal energy production may also be of interest. One component of energy metabolism is the ATPase 6 gene of mitochondria. In one Australian, a small deletion of two base pairs changes the last 25 amino acids in the carboxy terminus of this protein. No termination codon is brought into the frame, but the end of this protein has apparently some radical substitutions in amino acid sequence, and the gene may even be processed correctly, since the individual was phenotypically normal. Basic information about molecular sequences which are compatible with normal human activity levels can therefore be obtained.

We also learn about forces which control mutation rates in the human gene pool. Mitochondria may be constantly bombarded by mutagens, both internally from free oxygen radicals, and externally from the cellular environment (UV light, heat, dietary carcinogens, etc.). Measurements of DNA repair in the nucleus suggest that many enzymes are needed to ensure fidelity of replication and proper genetic function. These enzymes may be lacking almost completely in mitochondria. So do societies who cook and store their food in different ways, wear tight clothes, get fewer or more dental X-rays, or reproduce at relatively later ages, contribute mutations to the human gene pool that can be detected in their mitochondrial genes at high frequencies?

There is a bias to the class of mutations seen most often in mitochondrial genes. Transition mutations (the A to G or C to T variety) occur more frequently than transversion mutations (all the other possible combinations when purines mutate to pyrimidines, like A to C, or pyrimidines to purines, like C to G). The 9:1 ratio of bias seen from sequencing in the hominoid apes was not evident from restriction mapping studies, either in my own study or in a large survey of the Japanese population. The question therefore arises whether humans could repair some of these mutations differentially, or whether restriction enzymes were not adequate for these inferences, as they represented sequences of unknown importance which sometimes have nonrandom placement in the genome.

In sequencing through many functional regions, a few transversion mutations are in fact found. One interesting case involved the large ribosomal gene. Mutations found elsewhere in this gene can confer antibiotic resistance, and it is of primary importance to the translation of mitochondrial proteins. Two different sequences from separate clones were found for this gene from one individual, and the mutation was a transversion mutation at position 2940, where A has changed to T for the light strand of this clone, and a T to T mismatch presumably occurs.

The transition mutation bias may be region-specific. The D loop, or major noncoding region responsible for replication of the molecule, evolves primarily by transition mutations and small length mutations. If one wished to study recent divergence of population, for example that of Siberians, Aleuts, and Central American Indians, D loop sequences would give the most information in terms of absolute number of events which could be counted. However, if the point were to compare chimpanzee and human mtDNAs, one might better concentrate on ribosomal gene transversions, which occur less frequently and therefore have a lower probability of parallelism.

Mismatch repair DNA systems are probably not functional in mitochondria, and most of the mutations may simply be due to tautomeric shifts which would normally be repaired before replication in DNA if it were in the nucleus. Estimates of human mutation rates, based on the neutral theory, will have to take into account the actual fixation of new mutations, where the contributions of recombination, repair, and function can be distinguished. Hot spots can be detected, and we reach a better understanding of why parallel mutations are produced. Our resolution of these mutations is one reason that we state that most mtDNA polymorphisms will not be associated with a specific racial group. We learn that it is necessary to pay attention to specific types of mutations in tracing phylogeny, because they carry more functional significance and occur at a lower frequency in the population at large.

Time scales for human origins

The available data on mitochondrial polymorphisms in the literature begin to show a cohesive picture. From earlier work, one made assumptions about equal rates of evolution, mechanisms producing mutations, and population structure. Laboratory techniques were refined, more was learned of the molecular biology of the system, and more populations studied. Also, computer programs for cladistic analysis became available to handle larger data sets. Table 4 presents some of these findings.

A picture emerges of human populations diverging some time in the Middle Pleistocene. Using the overall rate of 2% per million years, these studies show that modern humans have an average time depth as a species of over 200,000 years. These studies used a variety of methods, and probed different areas of the world. All ethnic groups, with the exception of native Americans, demonstrate this pattern so far, and sub-Saharan Africans show roughly twice as much mitochondrial diversity as the rest of the world's population. To understand the significance of this finding, one considers what paleoanthropologists have said about the origin of anatomically modern humans. While a time depth of over 100,000 years can be documented by the fossil evidence in Africa (Rightmire, 1981), mitochondrial molecular genetics suggests that the gene pool for anatomically modern humans is at least twice as old as that estimate. In this Middle Pleistocene time range, and probably somewhere in sub-Saharan Africa (where there occurs the highest level of diversity indicative of the oldest continuous population centres), lie the roots of human genetic diversity.

Table 4: Estimates of human mtDNA diversity

Author	Type of data	Average % sequence divergence or $\delta/1000$	Time scale for common mother (years, b.p.)
Cann & Wilson, 1985	DNA sequence coding regions 8 kb 8 individuals	0.43	215,000
Loeb & Monat, 1985	DNA sequence coding regions 49 kb (repetitive clones) 5 individuals	0.5	250,000
Horai et al, 1984	Restriction maps purified mtDNA 62 sites 120 individuals	0.42	210,000
Greenberg et al, 1983	DNA sequence non-coding regions 6.3 kb 7 individuals	0.93	465,000* or 232,000
Johnson et al, 1983	Southerns of genomic DNA using mitochondrial probes 56 sites 235 individuals		220,000
Cann et al, 1982	Restriction maps purified mtDNA 441 sites 112 individuals	0.4	500,000
Brown, 1980	Restriction fragments purified mtDNA >200 fragments 21 individuals	0.36	180,000

*non-coding mtDNA estimate should be adjusted to *at least* a 4%/myr rate of evolution, double that measured in coding sequence and used here as 2%/myr. (From Brown et al, 1982.)

The constriction of human mitochondrial variability, relative to that measured in other species, suggests that some event may have triggered the dispersal or survival of a limited group of humans, and that the females from this group gave rise to all modern populations alive today. This conclusion profoundly affects the way one views the Middle Pleistocene hominids of Europe and Asia. Work on human Y chromosome sequences may allow us to test whether we can detect an ancestral population of males, tracing back to roughly the same time as that for these earth mothers. It will be of interest to see whether these males and the females converge on the same continent.

ACKNOWLEDGMENTS

I would like to thank W. Brown, F. Clark Howell, V. Sarich, M. Stoneking, T. White, and A.C. Wilson for spirited discussion of the topics covered here. This work was done in the laboratory of Professor Allan C. Wilson, Department of Biochemisty, University of California, Berkeley, 94720, with the help of many hands and hearts, and grant support from the National Science Foundation and Foundation for Research into the Origin of Man.

REFERENCES

Anderson, S., Bankier, A.T., Barrell, B.G., DeBruijn, M.H.C., Coulson, A.R., Drouin, J., Eperon, I.C., Nierlich, D.P., Roe, B.A., Sanger, F., Schreier, P.H., Smith, A.J.H., Staden, R. & Young, I.G. (1981). Sequence and organization of the human mitochondrial genome. Nature, **290**, 457-465.

Brown, W. (1980). Polymorphism in mitochondrial DNA of humans as revealed by restriction endonuclease analysis. Proceedings of the National Academy of Sciences, U.S.A., **77**, 3605-3609.

Brown, W. (1983). Evolution of animal mitochondrial DNA. In: M. Nei & R. Koehn (eds.), Evolution of Genes and Proteins, pp. 62-68. Sunderland, Mass: Sinauer Ass. Inc.

Brown, W., George, M. Jr. & Wilson, A.C. (1979). Rapid evolution of animal mitochondrial DNA. Proceedings of the National Academy of Sciences, U.S.A., **76**, 1967-1971.

Brown, W., Prager, E., Wang, A. & Wilson, A.C. (1982). Mitochondrial DNA sequences of primates: tempo and mode of evolution. Journal of Molecular Evolution, **18**, 225-239.

Cann, R. (1985) Mitochondrial DNA variation and the spread of modern populations. In: R. Kirk & E. Szathmary (eds.), Out of Asia - peopling the Americas and the Pacific, pp. 113-122. Canberra: Australian National University.

Cann, R. & Wilson, A. C. (1983). Length mutations in human mitochondrial DNA. Genetics, **104**, 699-711.

Cann, R., Brown, W. & Wilson, A.C. (1984). Polymorphic sites and the mechanism of evolution in human mitochondrial DNA. Genetics, **106**, 479-499.

Chomyn, A., Mariottini, P., Cleeter, M. Ragan, C.I., Matsuno-Yagi, A., Hatefi, Y., Doolittle, R., & Attardi, G. (1985). Six unidentified reading frames of human mitochondrial DNA encode components of the respiratory-chain NADH dehydrogenase. Nature, **314**, 592-597.

Giles, R., Blanc, H., Cann, H. & Wallace, D. C. (1980). Maternal inheritance of human mitochondrial DNA. Proceedings of the National Academy of Sciences, U.S.A., **77**, 6715-6719.

Greenberg, B., Newbold, J. & Sugino, A. (1983). Intraspecific nucleotide sequence variability surrounding the origin of replication in human mitochondrial DNA. Gene, **21**, 33-49.

Horai, S., Gojobori, T. & Matsunaga, E. (1984). Mitochondrial DNA polymorphism in Japanese. Human Genetics, **68**, 324-332.

Johnson, M., Wallace, D., Ferris, S., Rattazzi, M. & Cavalli-Sfrorza L. L. (1983). Radiation of human mitochondrial DNA types analyzed by restriction endonuclease cleavage patterns. Journal of Molecular Evolution, **19**, 255-271.

Messing, J. (1983). New M13 vectors for cloning. Methods in Enzymology, **101**, 20-78.

Monnat, R. & Loeb, L. (1985). Nucleotide sequence preservation of human mitochondrial DNA. Proceedings of the National Academy of Sciences, U.S.A., **82**, 2895-2899.

Oliver, N., Greenberg, B. & Wallace, D. (1983). Assignment of a polymorphic polypeptide to the human mitochondrial DNA unidentified reading frame 3 gene by a new peptide mapping strategy. Journal of Biological Chemistry, **258**, 5834-5839.

Rightmire, R. (1981). Later Pleistocene hominids of eastern and southern Africa. Anthropologie, **19**, 15-26.

Wilson, A. C., Cann, R., Carr, S., George, M., Gyllensten, U., Helm-Bychowski, K., Higuchi, R., Palumbi, S., Prager, E., Sage, R. & Stoneking, M. (1985). Mitochondrial DNA and two perspectives on evolutionary genetics. Biological Journal of the Linnean Society, **26**, 375-400.

MITOCHONDRIAL DNA VARIATION IN EASTERN HIGHLANDERS OF PAPUA NEW GUINEA

M. STONEKING[1,2], K. BHATIA[3] and A. C. WILSON[1]

Departments of [1]Biochemistry and [2]Genetics, University of California, Berkeley, California, U.S.A.

[3]Papua New Guinea Institute of Medical Research, Goroka, Eastern Highlands Province, Papua New Guinea.

INTRODUCTION

Studies of genetic variation in humans have contributed much to knowledge of the history and evolution of our species (Mourant et al, 1976; Nei & Roychoudhury, 1982; Ammerman & Cavalli-Sforza, 1985; Nei, 1985), providing a quantitative and temporal framework that is generally unavailable from other types of evidence. This is particularly true for studies of human population structure and migration in the South Pacific (Kirk, 1980a, 1982; Serjeantson, 1985). However, the variables typically employed in such studies (blood group and histocompatibility antigens, serum proteins, red cell enzymes, etc.) are inherited from both parents, and their recombination further obscures already complex patterns of variation that reflect historical interactions between such forces as mutation, selection, drift, migration, invasion, and extinction. Thus multiple interpretations are possible of the patterns of variation observed in these products of nuclear genes. To circumvent some of these problems, an alternative approach is to study restriction site polymorphisms in human mitochondrial DNA (mtDNA).

There are a number of advantages to studying mtDNA. The molecule is circular and small, about 16,500 base pairs (bp), and present in high copy number, so that it can be readily purified and analyzed by mapping with restriction enzymes. It is perhaps the best-known piece of eukaryotic DNA, with the complete nucleotide sequence and gene organization known from five species, including a human (Anderson et al, 1981). Knowledge of one human sequence facilitates high-resolution mapping of restriction enzyme sites in other human mtDNAs. This permits study of the nature and location of nucleotide substitutions and length mutations that have accumulated in the mtDNA molecule (Cann & Wilson, 1983; Cann et al, 1984). MtDNA evolves 5-10 times faster than nuclear DNA (Brown et al, 1979), with numerous differences accumulating between individuals within the same population (Brown, 1980;

Johnson et al, 1983; Horai et al, 1984; Cann, 1985; Wallace et al, 1985). It is strictly maternally inherited with no apparent recombination (Giles et al, 1980). This last feature is of particular importance; it allows one to identify maternal lineages and matriarchal relationships within a population, without the complications associated with sexual inheritance. Furthermore, the mutational divergence between two individual mtDNA lineages appears to accumulate in a regular clock-like fashion, permitting estimates of time elapsed since divergence from a common, ancestral mother. In contrast, gene frequencies at loci encoded in the nucleus are not expected to change in a simple, time-dependent manner. Fuller discussion of these and other aspects of mtDNA evolution is provided in recent reviews by Avise and Lansman (1983), Brown (1983), and Wilson et al (1985).

This paper is a preliminary account of the results of a survey of restriction site variation in mtDNA from human populations in New Guinea. Linguistically, culturally, and genetically this is one of the most diverse regions in the world, and has been the subject of a variety of intensive investigations (Wurm et al, 1975; Kirk, 1980b; Lin et al, 1983; Pietrusewsky, 1983; Serjeantson et al, 1983; Wurm, 1983). We present here an additional perspective, obtained from results on mtDNA polymorphisms in the Eastern Highlands of Papua New Guinea.

MATERIALS AND METHODS

Purification of mtDNA

MtDNA was purified from placental tissue by differential ultracentrifugation through cesium chloride density gradients (Cann, 1982). Purified mtDNA was digested with 13 restriction enzymes and the resulting fragments were end-labelled with radioactive nucleotides (Brown, 1980). The fragments were then separated by size by gel electrophoresis and visualised by autoradiography. Both agarose and polyacrylamide gels were used at a variety of concentrations so that all of the fragments from a particular digest could be visualised. For example, the restriction enzyme *Hae*III cuts human mtDNA into about 50 fragments ranging in size from 1611 bp to 15 bp; gels consisting of 2% agarose, 5% acrylamide or 8% acrylamide were used to resolve all of these fragments.

Restriction mapping

Differences in restriction fragment patterns were mapped by the sequence comparison method. By comparing the observed fragment patterns with the pattern predicted by the published reference sequence (Anderson et al, 1981), it was possible to map the precise location where the substitution responsible for changing the fragment pattern occurred. Additionally, if the change was a gain of a restriction site not present in the published sequence, then the exact substitution could be inferred. This was done by assuming that a single nucleotide substitution is responsible for an observed site change. One then searches for semisites (i.e., sites that differ from the restriction enzyme recognition sequence by one substitution) that occur in the right place to generate two fragments of the length observed. For example, Figure 1 illustrates a variant *Hinc*II restriction fragment pattern, found only in New Guinea mtDNA. The variant mtDNA lacks a 3939 bp fragment, and has in its

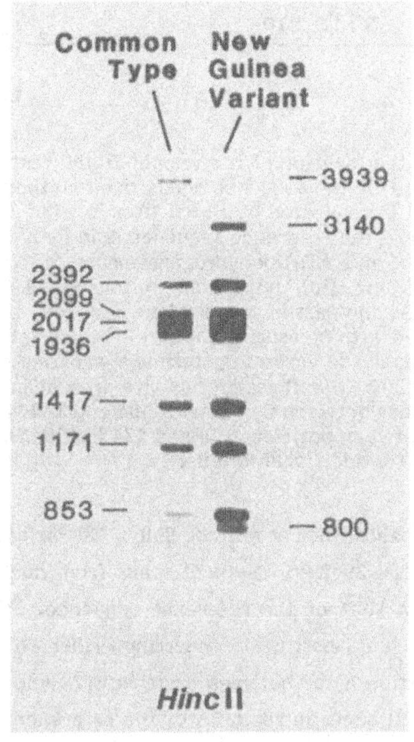

Figure 1: *Hinc*II fragments from two New Guinea mtDNAs, separated by electrophoresis through a 1.2% agarose gel. A single base change accounts for the differences between the two fragment patterns: the variant pattern (right) lacks a 3939 bp band present in the common pattern (left) and has, in its place, two new bands of about 3140 bp and 800 bp. The variant pattern was found in 12 of the 26 New Guinea mtDNAs; it has not been observed in over 500 human mtDNAs from outside New Guinea (Johnson et al, 1983; Cann et al, 1984; Horai et al, 1984; Wallace et al, 1985).

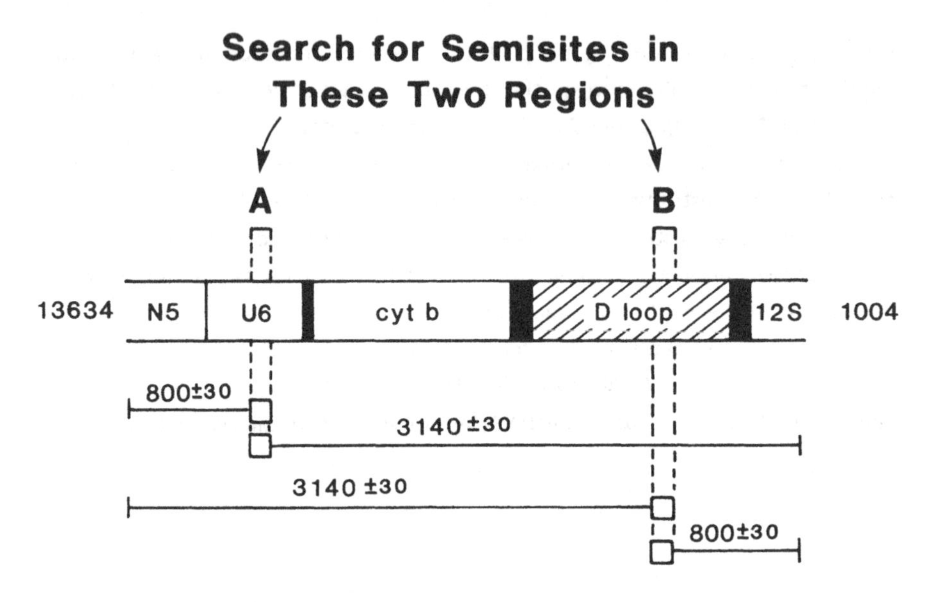

Figure 2: Mapping the location of the variant *Hinc*II restriction site depicted in Figure 1 by the sequence comparison method. The 3939 bp fragment is generated by *Hinc*II sites located at positions 13634 and 1004 of the reference sequence, and includes coding sequences for part of an NADH dehydrogenase subunit (N5), an unidentified reading frame (U6), cytochrome b, four tRNAs (solid shading), 12S rRNA, and part of the displacement loop (D loop). The error associated with measuring the size of the smaller of the two new bands in the variant pattern is approximately 30 bp, so that the additional *Hinc*II restriction site present in the variant pattern maps either between positions 14404 and 14464 (region A, in U6) or between positions 174 and 234 (region B, in the D loop). See text for additional details.

place two additional bands estimated to be 800 ± 30 bp and 3140 ± 30 bp. As Figure 2 illustrates, the 3939 bp fragment spans from nucleotide position 13634 to nucleotide position 1004 of the reference sequence. The variant pattern is thus due to a nucleotide substitution occurring either between positions 14404 and 14464 (Fig. 2, region A) or between positions 174 and 234 (Fig. 2, region B). No semisites for *Hinc*II occur in region A of the reference sequence, but there is one in region B, beginning at position 207:

Reference Sequence:	^{207}GTTAA<u>T</u>^{212}
New Guinea Variant:	GTTAA<u>C</u>

A transition (T → C) at position 212, which is within the displacement loop, can thus account for this variant *Hinc*II restriction pattern.

Table 1: Restriction enzymes used, recognition sequences, number of sites recognised in the published human mtDNA sequence, and number of restriction sites polymorphic in New Guinea

Enzyme	Recognition sequence*	No. of sites	No. of polymorphic restriction sites
HpaI	GTTAAC	3	1
FnudII	CGCG	6	0
AvaII	GGXCC	8	2
HincII	GTYRAC	12	1
HhaI	GCGC	17	3
HpaII	CCGG	23	1
MboI	GATC	23	1
TaqI	TCGA	29	3
RsaI	GTAC	35	8
HinfI	GANTC	36	4
HaeIII	GGCC	50	5
AluI	AGCT	64	6
DdeI	CTNAG	72	3

* A = adenine; T = thymine; G = guanine; C = cytosine; Y = pyrimidine (T or C); R = purine (A or G); X = A or T; N = any base.

Phylogenetic analysis

Phylogenetic trees depicting genealogical relationships of the mtDNA lineages were constructed using the computer program PAUP, developed by Dr. D. Swofford. The program employs a maximum parsimony algorithm to determine a tree of minimal length, using the presence or absence of restriction sites as character state data. Only phylogenetically informative sites (i.e., sites present in at least two and absent in at least two individuals) were used to construct the tree. Maximum likelihood estimates of pairwise sequence divergence were calculated by the iterative method of Nei and Tajima (1983). Time scales were derived by assuming that mtDNA sequence divergence accumulates at a constant rate of 2% per million years (Wilson et al, 1985).

RESULTS AND DISCUSSION

With the high-resolution mapping technique purified mtDNA was analysed from 24 individuals from the Eastern Highlands of Papua New Guinea, and from one individual each from the Southern Highlands and Morobe provinces. The 13 restriction enzymes used are listed in Table 1, along with recognition sequences, the number of sites in the published human mtDNA sequence, and the number of polymorphisms found. These 13 restriction enzymes recognised an average of about 370 sites per individual, which in turn represents about 1500 nucleotides, or 9% of the total human mtDNA genome.

Polymorphic Sites

Seventeen different types of mtDNA were found among the 26 New Guineans examined. Thirty-eight restriction sites were polymorphic, i.e. absent in at least one individual. In three cases it appears that two different restriction enzymes detect polymorphisms due to the same single-base change. *HpaI* cleaves a subset of *HincII* sequences; the *HincII* site gain illustrated in Figures 1 and 2 is also a *HpaI* site again. A G → A substitution at position 8251 would result in simultaneous loss of a *HaeIII* site and gain of an *AvaII* site, while a C → A substitution at position 5987 would result in simultaneous loss of a *HinfI*

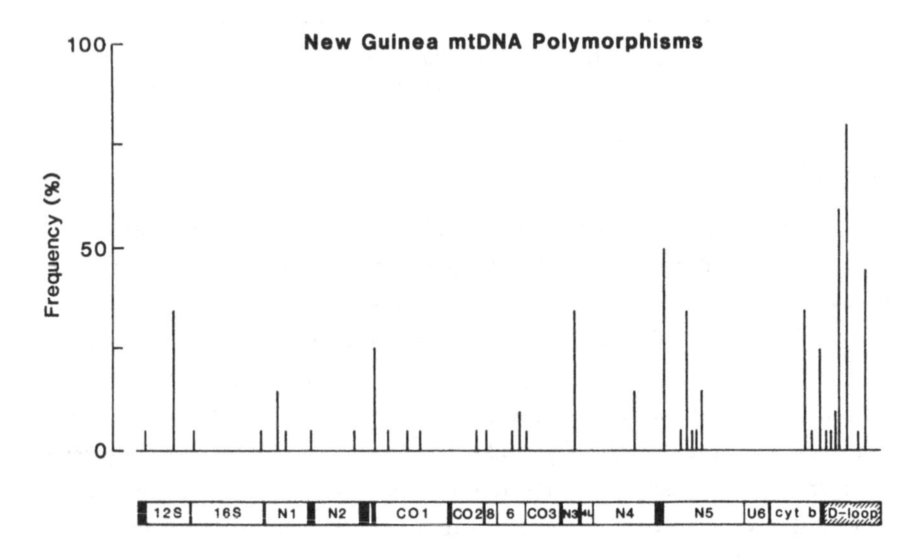

Figure 3: Distribution and frequency of 35 polymorphic sites found in 26 New Guinea mtDNAs. Along the bottom is a linear representation of the circular mtDNA molecule, starting from an arbitrary point (at the beginning of the phenylalanine tRNA gene). The regions of known function include: 22 tRNA genes (solid shading); two rRNA genes (12S and 16S); 13 protein-coding genes, including six NADH dehydrogenase subunits (N1-N5 and 4L), three cytochrome oxidase subunits (CO1-3), two ATPase subunits (6 and 8), cytochrome b, and one unidentified reading frame (U6); and the displacement loop (D loop), indicated by diagonal shading. Vertical lines indicate the location of each polymorphic site in the genome and its frequency in the New Guinea population.

site and gain of an *RsaI* site. Hence, 35 variable base-pair positions may account for these 38 polymorphic restriction sites; in the following analyses these 35 variable positions are referred to as *polymorphic sites*.

Figure 3 gives the frequency of each polymorphic site and its location in the mtDNA genome. The distribution of polymorphic sites appears to be random across the genome, with the exception of a cluster of seven sites in the displacement loop. This is consistent with previous studies showing that this noncoding region is the most variable part of human mtDNA (Aquadro & Greenberg, 1983; Cann et al, 1984).

Twenty-five of the polymorphic sites observed in New Guinea were private polymorphisms; that is, they have not been observed in mtDNA analyzed from an additional 121 humans of African, Asian, European, or aboriginal Australian descent (R. Cann, M. Stoneking, and A. C. Wilson, unpublished results). Some of these private polymorphisms reach appreciable frequencies, consistent with the remote, isolated nature of the New Guinea Highlands area.

Restriction Site Gains

Eighteen of the 35 polymorphic sites represent site gains, relative to the reference sequence, and hence the substitution can be inferred (Table 2). Eighty percent of these (14 out of 18) are transitions (A<->G or T<->C substitutions). Nine site gains occur in protein-coding regions, with four causing amino acid substitutions. These results are generally in good agreement with previous studies of restriction site variation in human mtDNA (Cann et al, 1984; Cann, 1985).

Phylogenetic Analysis

Figure 4 depicts a phylogenetic tree of minimal length relating the 26 New Guinea mtDNAs. Since mtDNA is strictly maternally inherited and does not undergo recombination, the phylogenetic tree also represents a genealogy of maternal lineages. This tree is based on 15 phylogenetically informative sites, and requires 17 mutations to account for the variation observed at these sites. The tree was rooted by the midpoint method, which places the common ancestor at the midpoint of the longest path connecting any two lineages, and thus assumes a constant rate of mtDNA sequence divergence.

The tree contains five primary branches (labelled 1-5), each leading to a cluster of one to 13 lineages. The time scale indicates that the common maternal ancestor of each cluster lived at times ranging from 5,000 to 60,000

Table 2: Location and nature of 18 site gains in 26 New Guinea mtDNAs

Location and base		Base change	Number of individuals	Amino acid substitution
D loop	65	T-C	1	-
	212	T-C	12	-
12S rRNA	751	A-C	1	-
	1404	A-G	9	-
tRNA^{trp}	5539	A-C	1	-
CO1	5987	C-A	7	silent
	6614	T-C	1	silent
	6917	G-A	1	silent
CO2	8251	G-A	1	silent
N3	10398	A-G	10	Thr-Ala
N4	11807	A-G	3	Thr-Ala
N5	12346	C-T	13	His-Tyr
	13071	C-T	1	silent
	13099	G-C	1	Ala-Pro
tRNA^{thr}	15900	T-C	1	-
D loop	16180	A-G	6	-
	16219	A-G	1	-
	16257	C-T	1	-

years ago. Fossil evidence for the presence of humans in New Guinea is scanty, but the earliest evidence dates back at least 30,000 years ago (White & O'Connell, 1982; Hope et al, 1983). Furthermore, humans were known to be in southern Australia at least 40,000 years ago (White & O'Connell, 1982; Wolpoff et al, 1984). Therefore, it is conceivable that occupation of the New Guinea Highlands began 50,000–60,000 years ago. The tree thus leads to the hypothesis that at least five maternal lineages colonised New Guinea, with subsequent diversification occurring within New Guinea.

If this hypothesis is correct, a phylogenetic treatment of these New Guinea mtDNA lineages along with other humans should reveal that: (a) lineages within the five clusters would still be associated; and (b) the five clusters themselves would each share a most recent common ancestor with a non-New Guinea mtDNA lineage. A phylogenetic tree of 134 mtDNA types, relating the 17 types of New Guinea mtDNA to 117 human mtDNAs from other

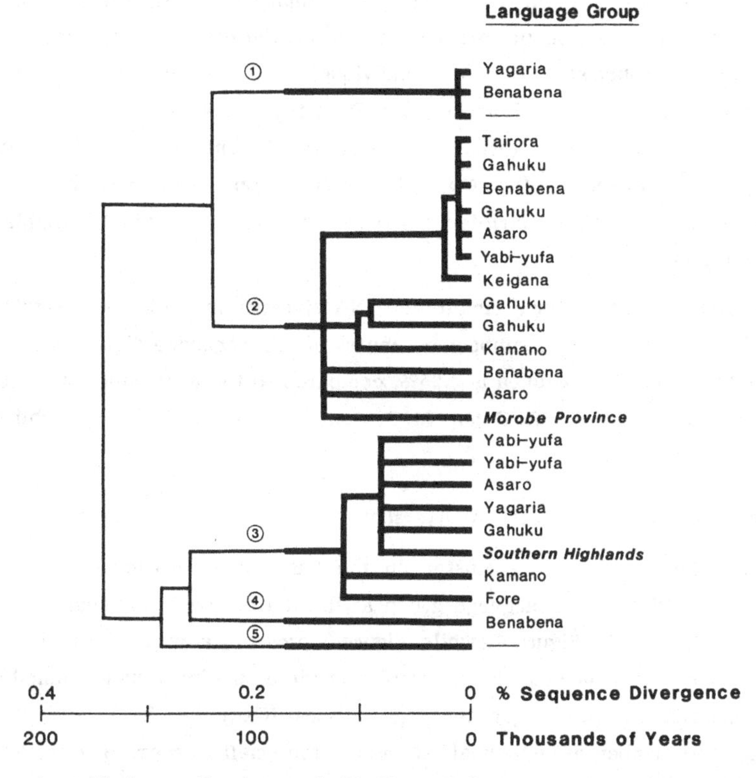

Figure 4: Phylogenetic tree of 26 New Guinea mtDNAs. The branching pattern was constructed by a maximum parsimony analysis of 15 phylogenetically informative sites; branch lengths were then adjusted to reflect maximum likelihood estimates of pairwise sequence divergence, averaged across nodes of the tree. The time scale was derived by assuming that human mtDNA sequence divergence accumulates at a constant rate of 2% per million years. The circled numbers and thicker lines indicate the five main branches. The language group affiliation of each individual is also indicated.

parts of the world (Cann et al, 1986), confirms this prediction. The five main clusters identified in Figure 4 remain intact but appear at different, widely-spaced locations in the tree for 134 mtDNA types.

The conclusion that the five primary branches in Figure 4 reflect five separate colonisations of the Highlands by maternal lineages rests, in part, on the assumption that the closest relative of each New Guinea lineage has already been found in our worldwide survey of human mtDNA's. Although this assumption is probably incorrect, it is unlikely that the discovery of additional mtDNA types will significantly alter the five primary branches. Three branches (1, 4, 5) each lead to a single mtDNA type, while the other two branches (2, 3) each lead to a subcluster of several individuals with identical mtDNA types. Even though additional non-New Guinea mtDNA types may be found to lie within one of the five primary clusters, the subclusters of identical lineages are very unlikely to be broken up further. For this reason, we expect the overall conclusion of a relatively restricted origin for mtDNA diversity in the Highlands to be maintained.

Further evidence for restricted mtDNA diversity in the Eastern Highlands comes from the maximum likelihood estimates of pair sequence divergence. The mean value within New Guinea is 0.25%, compared to the worldwide estimate of 0.36% (Brown, 1980; R. Cann, M. Stoneking & A. C. Wilson, unpublished results).

Language, Geography and mtDNA Diversity

Do the mtDNA types cluster on the basis of linguistic affiliation or geographic origin? The language group affiliation of each individual is shown alongside the tree in Figure 4, while Figure 5 presents a map of the Eastern Highlands Province, showing the geographic origin of the individuals studied and the boundaries of the major language groups (Gadjusek & Alpers, 1972). Although the number of individuals studied is too small to reach any definitive conclusions, there is no obvious association of mtDNA type with language group or geographic origin. This result is perhaps to be expected, since females in the new Guinea highlands are likely to disperse from their natal origin, without appreciable linguistic constraints (Wood et al, 1985).

CONCLUSION AND PROSPECTS

This analysis of mtDNA restriction site polymorphisms from the Eastern Highlands revealed: (a) high frequencies of private polymorphisms; (b) relatively low diversity; and (c) few major lineages, as defined by phylogenetic tree analysis. These results are consistent with a restricted origin for the occupation of the Highlands, with the colonizing population maintained in relative isolation. Furthermore, the lack of association between mtDNA types and language groups is paralleled by a relative lack of association between genetic polymorphisms

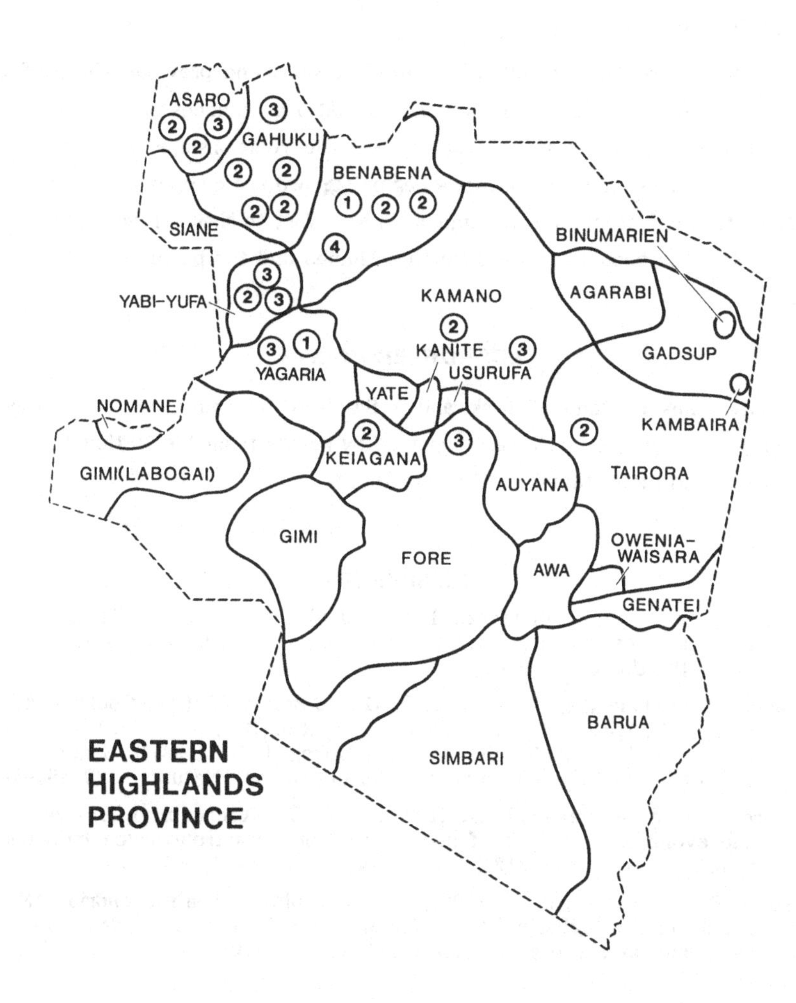

Figure 5: Map of the language groups of the Eastern Highlands, showing the origin of 22 of the 26 New Guinea mtDNAs (two were from outside the Eastern Highlands and two were of unknown origin from within the Highlands). Circled numbers, indicating the location of the village of origin, correspond to the five main branches identified in the tree presented in Figure 4.

constitute a single broad, genetic unit.

The correspondence between our conclusions and previous views of the origin and structure of Eastern Highland populations enhances our confidence in the utility of mtDNA for investigating genetic relationships between human populations. This study represents a beginning; we are currently extending our analyses to other areas of New Guinea, in an effort to shed further light on the spread of human populations through New Guinea and the peopling of the South Pacific.

ACKNOWLEDGMENTS

We thank R. Cann, R. Kirk, and D. Swofford for help in various phases of this project. The research was supported by grants from the National Science Foundation and the National Institutes of Health.

REFERENCES

Ammerman, A. J. & Cavalli-Sforza, L. L. (1985). The Neolithic Transition and the Genetics of Populations in Europe. Princeton, New Jersey: Princeton University Press.

Anderson, S., Bankier, A. T., Barrell, B. G., de Bruijn, M.H.L., Coulson, A. R., Drouin, J., Eperon, I. C., Nierlich, D. P., Roe, B. A., Sanger, F., Schreier, P. H., Smith, A. J. H., Staden, R. & Young, I. G. (1981). Sequence and organization of the human mitochondrial genome. Nature, **290**, 457-465.

Aquadro, C. F. & Greenberg, B. D. (1983). Human mitochondrial DNA variation and evolution: analysis of nucleotide sequences from seven individuals. Genetics, **103**, 287-312.

Avise, J. C. & Lansman, R. A. (1983). Polymorphism of mitochondrial DNA in populations of higher animals. In: M. Nei & R. K. Koehn (eds.), Evolution of Genes and Proteins, pp. 147-164. Sunderland, Mass: Sinauer Associates.

Brown, W. M. (1980). Polymorphism in mitochondrial DNA of humans as revealed by restriction endonuclease analysis. Proceedings of the National Academy of Sciences, U.S.A., **77**, 3605-3609.

Brown, W. M. (1983). Evolution of animal mitochondrial DNA. In: M. Nei & R.K. Koehn (eds.), Evolution of Genes and Proteins, pp. 62-68. Sunderland, Mass: Sinauer Associates.

Brown, W. M., George, M. & Wilson, A. C. (1979). Rapid evolution of animal mitochondrial DNA. Proceedings of the National Academy of Sciences, U.S.A., **76**, 1967-1971.

Cann, R. L. (1982). The evolution of human mitochondrial DNA. Ph.D. thesis, The University of California, Berkeley.

Cann, R. L. (1985). Mitochondrial DNA variation and the spread of modern populations. In: R. Kirk & E. Szathmary (eds.), Out of Asia - Peopling the Americas and the Pacific, pp. 113-122. Canberra: Journal of Pacific History, Australian National University.

Cann, R. L. & Wilson, A. C. (1983). Length mutations in human mitochondrial DNA. Genetics, **104**, 699-711.

Cann, R. L., Brown, W. M. & Wilson, A. C. (1984). Polymorphic sites and the mechanism of evolution in human mitochondrial DNA. Genetics, **106**, 479-499.

Cann, R. L., Stoneking, M. & Wilson, A. C. (1986). Mitochondrial DNA and human evolution. Nature (in press).

Gadjusek, D. C. & Alpers, M. (1972). Genetics studies in relation to Kuru. American Journal of Human Genetics, **24**, S1-S110.

Giles, R. E., Blanc, H., Cann, H. M. & Wallace, D. C. (1980). Maternal inheritance of human mitochondrial DNA. Proceedings of the National Academy of Sciences, U.S.A., **77**, 6715-6719.

Hope, G. S., Golson, J. & Allen, J. (1983). Paleoecology and prehistory in New Guinea. Journal of Human Evolution, **12**, 37-60.

Horai, S., Gojobori, T. & Matsunaga, E. (1984). Mitochondrial DNA polymorphism in Japanese. Human Genetics, **68**, 324-332.

Johnson, M. J., Wallace, D. C., Ferris, S. D., Rattazzi, M. C. & Cavalli-Sforza, L. L. (1983). Radiation of human mitochondria DNA types analyzed by restriction endonuclease cleavage patterns. Journal of Molecular Evolution, **19**, 255-271.

Kirk, R. L. (1980a). Language, genes, and people in the Pacific. In: A.W. Eriksson (ed.), Population Structure and Genetic Disorders, pp. 113-137. New York: Academic Press.

Kirk, R. L. (1980b). Linguistic, ecological, and genetic differentiation in New Guinea and the Western Pacific. In: M.H. Crawford & J.H. Mielke (eds.), Current Developments in Anthropological Genetics, Vol. 2, Ecology and Population Structure, pp. 229-253. New York: Plenum Press.

Kirk, R. L. (1982). Microevolution and migration in the Pacific. In: B. Bonne-Tamir, P. Cohen & R. M. Goodman (eds.), Human Genetics, Part A: The Unfolding Genome, pp. 215-225. New York: Alan R. Liss.

Lin, P. M., Enciso, V. B. & Crawford, M. H. (1983). Dermatoglyphic inter- and intrapopulation variation among indigenous New Guinea groups. Journal of Human Evolution, **12**, 103-123.

Mourant, A. E., Kopec, A. C. & Domaniewska-Sobczak, K. (1976). The Distribution of the Human Blood Groups and Other Polymorphisms, 2nd Edition. Oxford: Oxford University Press.

Mourant, A. E., Tills, D., Kopec, A. C., Warlow, A., Teesdale, P., Booth, P. B. & Hornabrook, R. W. (1982). Red cell antigen, serum protein, and red cell enzyme polymorphisms in Eastern Highlanders of New Guinea. Human Heredity, **32**, 374-384.

Nei, M. (1985). Human evolution at the molecular level. In: K. Aoki & T. Ohta (eds.), Population Genetics and Molecular Evolution, pp. 41-64. New York: Springer-Verlag.

Nei, M. & Roychoudhury, A. K. (1982). Genetic relationship and evolution of human races. Evolutionary Biology, **14**, 1-59.

Nei, M. & Tajima, F. (1983). Maximum likelihood estimation of the number of nucleotide substitutions from restriction sites data. Genetics, **105**, 207-217.

Pietrusewsky, M. (1983). Multivariate analysis of New Guinea and Melanesian skulls: a review. Journal of Human Evolution, **12**, 61-76.

Serjeantson, S. W. (1985). Migration and admixture in the Pacific - insights provided by human leucocyte antigens. In: R. Kirk & E. Szathmary (eds.), Out of Asia - Peopling the Americas and the Pacific, pp. 133-145. Canberra: Journal of Pacific History, Inc., Australian National University.

Serjeantson, S. W., Kirk, R. L. & Booth, P. B. (1983). Linguistic and genetic differentiation in New Guinea. Journal of Human Evolution, **12**, 77-92.

Wallace, D. C., Garrison, K. & Knowler, W. C. (1985). Dramatic founder effects in Amerindian mitochondrial DNAs. American Journal of Physical Anthropology, **68**, 149-155.

White, J. P. & O'Connell, J. F. (1982). A Prehistory of Australia, New Guinea and Sahul. New York: Academic Press.

Wilson, A. C., Cann, R. L., Carr, S. M., George, M., Gyllensten, U. B., Helm-Bychowski, K. M., Higuchi, R. G., Palumbi, S. R., Prager, E. M., Sage, R. D. & Stoneking, M. (1985). Mitochondrial DNA and two perspectives on evolutionary genetics. Biological Journal of the Linnean Society, **26**, 375-400.

Wolpoff, M. H., Zhi, W. X. & Thorne, A. G. (1984). Modern Homo sapiens origins: a general theory of hominid evolution involving the fossil evidence from East Asia. In: F.H. Smith & F. Spencer (eds.), The Origins of Modern Humans: A World Survey of the Fossil Evidence, pp. 411-483. New York: Alan R. Liss.

Wood, J. W., Smouse, P. E. & Long, J. C. (1985). Sex-specific dispersal patterns in two human populations of highland New Guinea. American Naturalist, **125**, 747-768.

Wurm, S. A. (1983). Linguistic prehistory in the New Guinea area. Journal of Human Evolution, **12**, 25-35.

Wurm, S. A., Laycock, D. C., Voorhoeve, C. L. & Dutton, T. E. (1975). Papuan linguistic prehistory and past language migrations in the New Guinea area. In: S.A. Wurm (ed.), New Guinea Area Languages and Language Study, Vol. 1, Papuan Languages and the New Guinea Linguistic Scene, pp. 935-960. Canberra: Australian National University.

THE CONTRIBUTION OF POLYMORPHISMS IN mtDNA TO POPULATION GENETIC STUDIES

B. BONNÉ-TAMIR[1] and L.L. CAVALLI-SFORZA[2]

[1]*Department of Human Genetics, Sackler School of Medicine, Ramat Aviv, Israel.*

[2]*Department of Genetics, Stanford University School of Medicine, Stanford, U.S.A.*

INTRODUCTION

The advent of the powerful recombinant DNA techniques has given new perspectives on the origin and expansion of genetic diversity within the human species. Studies at the DNA level have yielded a host of new genetic markers and have thus greatly increased the number of human genetic differences that can be detected both among individuals as well as among populations. Genetic polymorphisms in the DNA itself can be identified either by direct sequencing of the DNA molecule - which is a laborious procedure requiring enormous effort - or by using restriction enzymes to cut the DNA at specific points and so to produce fragments whose lengths indicate whether or not the relevant base pair sequence is present in the DNA of the individual. Mitochondrial DNA which is a cytoplasmic gene system has in the last few years been very well characterized at the molecular level; possibly it is "the best known piece of human DNA" (Cann, 1982).

Most of the studies bearing on mtDNA variation in higher animals and humans have focused on evolutionary issues, emphasizing its uses as a marker of evolutionary phenomena (Brown, 1983); these included tracing patterns of common ancestry among species (Brown et al, 1982) and among human ethnic groups (Johnson et al, 1983), estimating times of divergence among major geographical groups of humans (Cann et al, 1982), and even attempting to unravel the evolutionary history of man and his relatives by comparisons of restriction maps of gorilla, orangutan, gibbon and chimpanzee (Ferris et al, 1981). But there are many more uses to which it can be put in understanding the biology of man.

The lack of variation of mtDNA types within individuals and its maternal inheritance make the mtDNA an excellent marker for the reconstruction of matriarchal lineages: each female transmits her own characteristic mtDNA molecule to all her progeny; hence all individuals from maternal lineages that are descendants of the same ancestral female are expected to have the same mtDNA type. For that reason it has been claimed that genealogical analysis may be simpler and more reliable for mtDNA than for nuclear genes (Avise et al, 1979).

This paper draws attention to the potential value of the mitochondrial genome for investigation of patterns of divergence between closely related populations, and provides thereby new insights on more subtle questions of ethno-historical relationships between sub-populations.

METHOD

By comparison with sequencing, the study of fragment patterns of electrophoresis of single enzyme digests is only an approximate method for detecting individual differences. Population studies, however, require the analysis of many individuals, and with current techniques there are serious limitations in practice to sequencing more than a few individuals. An alternative to the Southern Blot method has been employed by Brown et al (1982) and Cann et al (1982) and consists of end-labelling purified mtDNA. With such a method one can easily use for digestion four base pair cutters which produce more fragments and increase the amount of information.

This benefit is counterbalanced by a cost: in order to purify mtDNA one must obtain large amounts of DNA, which for humans, in practice means placental samples. This is a serious limitation, which inevitably decreases the number of individuals that can be tested and the populations on which this approach is possible. However the alternative, to use blood cell DNA, is also useful, as shown here. The two methods of analysis of restriction fragments by Southern Blots and by end-labelling serve, therefore, somewhat different functions and the choice between them depends on the particular population and problem.

The method employed, using total blood cell DNA, is as follows:

DNA isolation: Total human DNA is extracted from the buffy coats (Kan et al, 1977) including an overnight dialysis against 10 mM Tris HCL (pH 7.5/1 mM EDTA following the first phenol extraction).

Restriction analysis: Two micrograms of total DNA are used for each digestion. The conditions and buffers are those recommended by the manufacturer. Samples are digested with a five-fold excess of enzyme for 5-6 h and then heated at 65°C for 10 min to inactivate the enzyme. A dye marker of bromocresol purple is added along with glycerol to a final concentration of 10% (vol/vol). The digested fragments are separated on horizontal agarose slab gels ranging from 0.8% to 1.8% depending on the expected size of the restriction fragments. A known size marker is used for determination of fragment size.

Gels are run overnight at 1.5 v/cm; the fragments are then transferred (Southern, 1975) to a Zetabind filter (AMF Cuno). Hybridization is performed in 5xSSPE (1xSSPE = 0.18 M Nacl, 10 mM NaH_2PO_4, 1 mM Na_2 EDTA, pH 7.0) at 42°C for 48 h using mtDNA purified from human tissue culture cells (Giles et al, 1980) as a probe, nick-translated (Rigby et al, 1977) to a specific activity of $1x10^8$ cpm/mg. Filters are washed for 2 h in 2xSSPE 0.1% SDS and fragments can be visualised after overnight autoradiography.

RESULTS

As an illustration of the possible application of mtDNA variation to issues of genetic diversity among different ethnic groups, residing in the same geographical region, the following preliminary results from a pilot study of mtDNA types in two Israeli populations are given (Bonné-Tamir et al, 1986).

The study consisted of a comparison of mtDNA polymorphisms in two groups living in Israel - Jews and Arabs. They differ in their ethnic and religious affiliations, but both belong to the same major ethnic division (Caucasian) and represent populations who emerged as distinct peoples, perhaps 5-10 thousand years ago. From each subject 20-30 ml whole blood was collected into ACD. Altogether 82 Israeli Jews and Arabs were included in the sample. The total blood cell DNA extracted from each specimen was digested with five enzymes: Hpa I, Bam HI, Hae II, Msp I and AVA II. The frequencies of the variants in each polymorphism are given in Table 1, together with representative frequencies in other major populations.

Of the five enzymes, four proved particularly useful and revealed mtDNA variability, including new morphs (Figs. 1, 2). Bam HI was found to be the least informative. Two different cleavage patterns were detected by using the Hpa I restriction endonuclease; morphs 2 and 3. This finding among Israeli Arabs of morph 3, known previously only in African populations (Denaro et al, 1981), is consistent with earlier studies reporting the presence of typical "African" markers such as the phenotypes Fy^{a-b-}, Rho (cDe) and Sutter antigens (Levene et al, 1976; Sandler et al, 1979). The other enzymes also reveal a slight tendency towards "African" characteristics in the Arab group, such as the lower

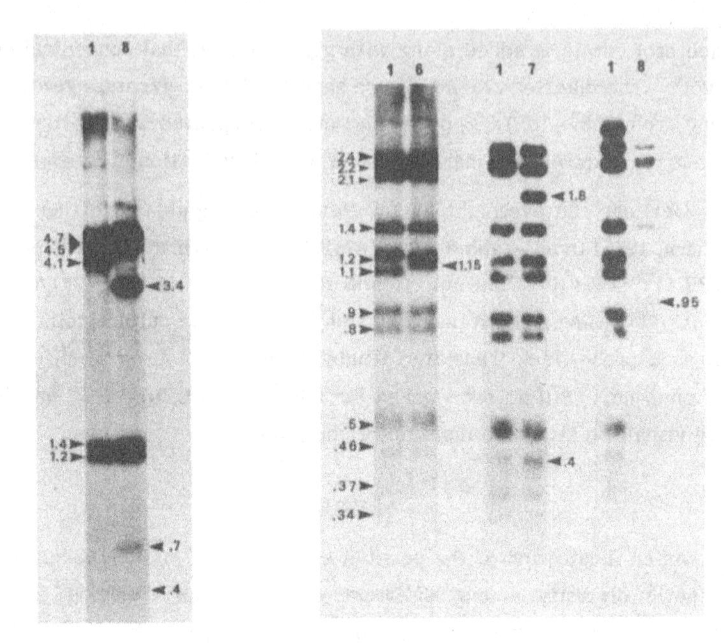

Figure 1: Autoradiograms of new Hae II and Msp I morphs found in this study, each aligned against morph 1 for comparison. Morphs are denoted by numbers at top of columns; fragment sizes are designated in kilobases.

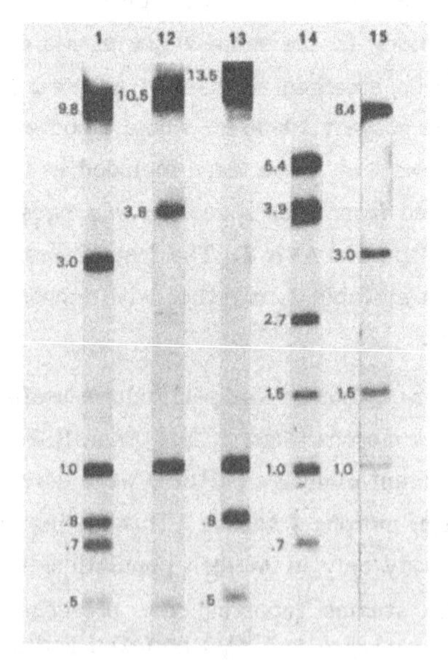

Figure 2: Autoradiogram of Ava II; four new morphs compared with morph 1 which corresponds to the published mtDNA sequence.

AVA II morph 1 frequency, the existence of AVA II types 3 and 5, and the lower frequency of morph 1 of the Msp 1 enzyme (Table 1).

When the restriction endonuclease morphs (fragment patterns) observed for each individual are combined, there were 18 distinct mtDNA

Table 1: Number of individuals and frequencies of all morphs listed by enzyme for the two Israeli groups compared with frequencies for three world populations(Denaro et al, 1981; Johnson et al, 1985)

Sample size		Israeli Jews		Israeli Arabs		Caucasians	Oriental	African (Bantu)
Morph		40		41		50	46	40
		No.	%	No.	%	%	%	%
Bam H1	-1	39	97.5	41	100.0	86.0	100.0	100.0
	-2	1	2.5	0	0.0	8.0	0.0	0.0
	-3	0	0.0	0	0.0	6.0	0.0	0.0
Hae II	-1	23	57.5	39	95.1	76.0	80.4	97.5
	-2	15	37.5	1	2.4	14.0	13.0	2.5
	-3	1	2.5	0	0.0	6.0	0.0	0.0
	-4	1	2.5	0	0.0	0.0	2.2	0.0
	-8	0	0.0	1	2.4	0.0	0.0	0.0
Hpa I*	-2	38	100.0	35	87.5	98.1	81.3	25.0
	-3	0	0.0	5	12.5	0.0	0.0	70.8
Ava II*	-1	29	76.3	27	69.2	74.0	95.7	40.0
	-2	0	0.0	0	0.0	4.0	0.0	12.5
	-3	-	0.0	2	5.1	2.0	0.0	37.5
	-4	0	0.0	0	0.0	0.0	0.0	0.0
	-5	5	13.1	8	20.5	8.0	0.0	5.0
	-9	1	2.6	0	0.0	6.0	0.0	0.0
	-12	0	0.0	1	2.6	0.0	0.0	0.0
	-13	1	2.6	1	2.6	0.0	0.0	0.0
	-14	1	2.6	0	0.0	0.0	0.0	0.0
	-15	1	2.6	0	0.0	0.0	0.0	0.0
Msp 1	-1	40	100.0	35	89.7	92.0	97.8	87.5
	-2	0	0.0	0	0.0	0.0	0.0	12.5
	-4	0	0.0	1	2.6	8.0	2.2	0.0
	-6	0	0.0	1	2.6	0.0	0.0	0.0
	-7	0	0.0	1	2.6	0.0	0.0	0.0
	-8	0	0.0	1	2.6	0.0	0.0	0.0

* The morph was not identified in all individuals.

Table 2: Frequencies of "New" and "Old" mtDNA types in the two Israeli groups compared with those found in three world populations (Denaro et al, 1981; Johnson et al, 1983)

	Enzyme Morph	Israeli Jews		Israeli Arabs		Caucasian	Oriental	African (Bantu)
		No.	%	No.	%	%	%	%
"Old" Type*								
1	(2-1-1-1-1-)	15	38.0	22	56.4	58.0	69.6	25.0
2	(3-1-1-1-3)	0	0.0	1	2.6	0.0	0.0	32.5
6	(2-1-2-1-1)	14	36.0	1	2.6	12.0	2.2	0.0
7	(3-1-1-1-1)	0	0.0	1	2.6	0.0	0.0	10.0
11	(2-2-3-1-5)	1	2.6	0	0.0	3.0	0.0	0.0
17	(2-1-1-1-9)	1	2.6	0	0.0	2.0	0.0	0.0
22	(2-1-1-1-5)	4	10.3	6	15.4	2.0	0.0	0.0
31	(3-1-1-1-5)	0	0.0	2	5.1	0.0	0.0	2.5
"New" Type								
36	(2-1-1-1-13)	1	2.6	0	0.0	0.0	0.0	0.0
37	(2-1-1-1-14)	1	2.6	0	0.0	0.0	0.0	0.0
38	(2-1-1-1-15)	1	2.6	0	0.0	0.0	0.0	0.0
39	(2-1-4-1-1)	1	2.6	0	0.0	0.0	0.0	0.0
40	(2-1-1-6-1)	0	0.0	1	2.6	0.0	0.0	0.0
41	(2-1-1-7-1)	0	0.0	1	2.6	0.0	0.0	0.0
42	(2-1-1-8-1)	0	0.0	1	2.6	0.0	0.0	0.0
43	(3-1-1-1-12)	0	0.0	1	2.6	0.0	0.0	0.0
44	(2-1-1-4-3)	0	0.0	1	2.6	0.0	0.0	0.0
45	(2-1-8-1-1)	0	0.0	1	2.6	0.0	0.0	0.0

"types" (Table 2). These include eight known fragment patterns (Johnson et al, 1983) but also ten new ones, which can be related to each other by single nucleotide substitutions (Figs. 2,3). Only three are shared by both Arabs and Israeli, and even those show striking frequency differences. Again there appear African combinations.

In spite of the very small numbers in each group the results demonstrate the existence of specific mtDNA fragment patterns varying in frequency from population to population, and suggest that certain types may be unique to certain groups.

Even with the limited use of only 5 restriction enzymes, a high correlation between mtDNA types and the ethnic origin of the individuals is confirmed. These preliminary findings are thus in accord with statements such as those made by Brown (1980) that shared mtDNA polymorphisms may indicate group affinities within major ethnic groups and that Caucasian, Oriental and African mtDNA types can be distinguished in those where these populations have contributed to the gene pool (Johnson et al, 1983).

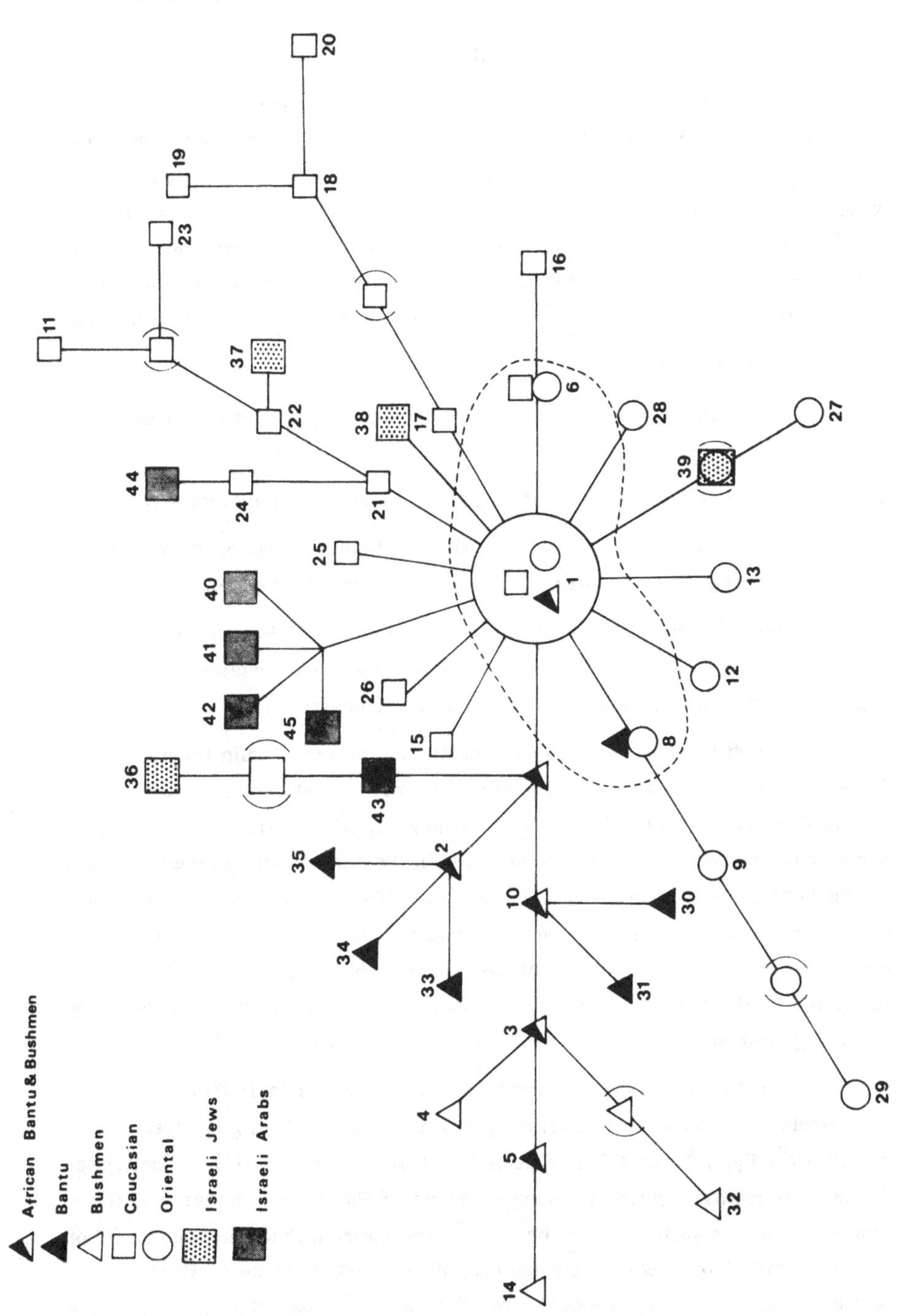

Figure 3: Phylogeny of mtDNA types including the 10 new types observed (Bonné-Tamir et al, 1986).

DISCUSSION

This study suggests that polymorphisms in mtDNA as revealed by restriction endonuclease analysis can yield an independent measure on life histories and degree of relatedness between sub-populations as well as account for and explain observations on gene frequencies of classical "nuclear" polymorphisms. It may seem almost paradoxical that such molecular genetic techniques can provide answers to problems of an anthropological-demographic nature, probably more accurate than historical evidence, written records and orally transmitted legend.

Several such problems are:

1. Genealogical reconstruction of lineages, including mating preferences and inbreeding patterns.
2. Extent of genetic isolation of communities and direction of gene flow.
3. Types of migrations; i.e. consisting of family units, or migrations of males only, such as hunters, travellers, soldiers, and the like.

As this type of investigation requires much time and financial investment, emphasis should be directed less to large scale descriptive characterization of populations, and more to concentration on unsolved questions.

One such topic is the origin of certain South-Sinai Beduin tribes. Some 12000 individuals constitute a confederation of several tribes known as the "Towara" (Bonne et al, 1971; Bonne-Tamir, 1981). Among them is one exceptional tribe, the "Jebeliya", who reside in the central part of the peninsula, in the vicinity of St. Catherine's Monastery. Opinions differ regarding their origin. Some claim that they are descendants of European serfs who were deported with their wives in the 6th century A.D. by Justinian and subsequently converted to the Moslem faith (Burckhardt, 1822). Others maintain that they are of Egyptian and Greek origin (Shocair, 1916; Oppenheim, 1941).

Preliminary results of serological tests carried out in 1967/68 pointed to considerable incorporation of African genes, as shown by high gene frequencies of cDe (R^o) Fy, Js^a and G6PD type A (Gd^a) (Bonne et al, 1971). Analysis of fragment patterns in mtDNA among members of South-Sinai tribes will throw light on questions such as were the women who married the "European serfs" of St. Catherine's Monastery of African origin? Or, despite some of their negroid features, how can one account for the obvious dissimilarity to African frequencies in many blood group antigens such as B, cde, P, Kidd and Fy^a?

Another subject worth exploration is the origin of the Ethiopian Jews known as the Falasha. Some scholars believe that they are indigenous Africans who were converted to Judaism possibly by migrants from Yemen. Yet, when an array of gene frequencies including bloodgroups and red cell enzymes are used in genetic distance measurements, cluster or kinship analysis, these two populations, one until recently residing in Saudi-Arabia, the other from north east Africa, appear genetically very distant from each other. Mitochondrial polymorphisms may help to clarify whether Yemenite missionaries, traders and others, who came to Ethiopia, actually played any part in the *biological* history of Ethiopian Jews.

MtDNA studies can thus prove valuable for checking theories about human history, especially theories about migrations, dispersions and admixture.

REFERENCES

Avise, J. C., Giblin-Davidson, C., Laerm, J., Patton, J. C. & Lansman, R. A. (1979). Mitochondrial DNA clones and matriarchal phylogeny within and among geographic populations of the pocket gopher *Geomys pinetis*. Proceedings of the National Academy of Sciences, U.S.A., **76**, 6694-6698.

Bonné, B., Godber, M., Ashbel, S., Mourant, A. E. & Tills, D. G. (1971). South-Sinai Beduin. A preliminary report on their inherited blood factors. American Journal of Physical Anthropology, **34**, 397-408.

Bonné-Tamir, B. (1981). Genetic markers and historical inference: a case study among South Sinai Beduin. International Symposium on Abnormal Hemoglobins. Genetics, Populations and Disease, Jerusalem (Abstract).

Bonné-Tamir, B., Johnson, M. J., Natali, A., Wallace, D. C. & Cavalli-Sforza, L. L. (1986). Human mitochondrial DNA types in two Israeli populations - a comparative study at the DNA level. American Journal of Human Genetics, **38**, 341-351.

Brown, W. M. (1983). Evolution of animal mitochondrial DNA. In: M. Nei and R. Koehn (eds.), Evolution of genes and protein, pp. 62-88. Sunderland, Mass: Sinauer.

Brown, W. M., Prager, E. M., Wang, A. & Wilson, A. C. (1982). Mitochondrial DNA sequences of primates: tempo and mode of evolution. Journal of Molecular Evolution, **18**, 225-239.

Burckhardt, J. L. (1822). Travels in Syria and the Holy-Land. London: Murray.

Cann, R. L., Brown, W. M. & Wilson, A. C. (1982). Evolution of human mitochondrial DNA: a preliminary Report. In: B. Bonne-Tamir, T. Cohen & R.M. Goodman (eds.), Human Genetics, Part A: The Unfolding Genome, p. 157. New York: Alan R. Liss.

Loghem, E. van, & Modiano, G. (1969). Studies on African pygmies, i. A pilot investigation of Babinga pygmies in the Central African Republic (with an analysis of genetic distances). American Journal of Human Genetics, **21**, 252-274.

Denaro M., Blanc, H., Johnson, M. J., Chen, K. H., Wilmsen E., Cavalli-Sforza, L. L. & Wallace D. C. (1981). Ethnic variation in Hpa I endonuclease cleavage patterns of human mitochondrial DNA. Proceedings of the National Academy of Science, U.S.A., **78**, 5768-5772.

Ferris, S. D., Wilson, A. C. & Brown, W. M. (1981). Evolutionary tree for apes and humans based on cleavage maps of mitochondrial DNA. Proceedings of the National Academy of Science, U.S.A., **78**, 2432-2436.

Giles, R. E., Blanc, H., Cann, H. M. & Wallace, D. C. (1980). Maternal inheritance of human mitochondrial DNA. Proceedings of the National Academy of Science, U.S.A., **77**, 6715-6719.

Johnson, M. J., Wallace, D. C., Ferris, S. D., Rattazzi, M. C. & Cavalli-Sforza, L. L. (1983). Radiation of human mitochondrial DNA types analyzed by restriction endonuclease cleavage patterns. Journal of Molecular Evolution, **19**, 255-271.

Kan, Y. W., Dozy, A. M., Trecartin, R. & Todd, D. (1977). Identification of a deletion defect in thalassaemia. New England Journal of Medicine, **297**, 1080-1084.

Levene, C., Rachmilewitz, E. A., Ezekiel, E., Freundlich, E. & Sandler, G. (1976). Blood group phenotypes and hemoglobin S. Acta Haematologica, **55**, 300-305.

Oppenheim, M. F. (1943). Die Beduinen in Palastina, Transjordanien, Sinai und Hedgas.

Rigby, P.W.J., Dickerman, M., Rhodec, C. and Berg, P. (1977). Labelling deoxyribonucleic acid to high specific activity *in vitro* by nick translation with DNA polymerase I. Journal of Molecular Biology, **113**, 237-251.

Shocair, N. (1916). The History of Sinai and the Arabs. Cairo (in Arabic).

Southern, E. M. (1975). Detection of specific sequences among DNA fragments separated by gel electrophoresis. Journal of Molecular Biology, **98**, 503-517.

PART II

GENETIC DIVERSITY - ITS ORIGIN AND MAINTENANCE

HUMAN GENETIC DIVERSITY IN SOUTH-EAST ASIA AND THE WESTERN PACIFIC

R. L. KIRK

Department of Human Biology, John Curtin School of Medical Research, Canberra, Australia.

INTRODUCTION

The scenarios discussed in this paper are based on my own work during the last 25 years, supplemented with studies of other investigators, and made possible through collaboration with many colleagues. To these persons, too numerous to mention here, I am deeply indebted.

Our investigations have been concerned with people of many different ethnic origins, covering a wide geographical area, from India and Sri Lanka in the west, through Malaysia, Thailand, Indonesia, and New Guinea, to the central Pacific in the east. Population structures of the peoples studied range also from small tribal enclaves within larger communities, or isolated island populations through to populations with well defined social castes of wide dispersal. The genetic variability is equally varied, and the genetic relationships between these populations is complex, being a compound of diversity of origins, migration and intermixture, and adaptation to particular environments. Certainly we do not yet fully understand this complexity, but we have unravelled some part of it, enough now to see what needs to be studied further.

These studies have covered human populations in the tropics. This does not mean that all live in a uniform environment. Their habitats vary from tropical rain-forest to desert, from sea level to several thousand metres elevation. Their economic base likewise covers hunter-gatherers, slash and burn cultivators, to settled agriculturalists and urban dwellers. Sometimes ecological differences are broadly based, as in India and South-East Asia, sometimes they occur within a small geographic area.

ECOLOGICAL VARIATION

India and South-East Asia

For India and South-East Asia there are broad ecological zones. In India there are bands of monsoon forest along the west coast, with deciduous monsoon forest on the slopes of the western Ghats and a large area of scrub woodland to the east, with a broad band of deciduous monsoon forest to the north. True monsoon forest occupies N.E. India and the entire west coast of S.E. Asia, with bands of deciduous monsoon forest and scrub woodland to the east (Bowles, 1977, p. 27).

Within these broad zones, however, there are localised variations which are of importance for the human populations. For example in S.W. India a transect from the west coast towards the east shows the effect of elevation. The wet western slopes are heavily forested, the higher plateaus of the Nilgiri Hills are open grasslands, with only small pockets of monsoon forest. Descending their eastern slopes one passes through a zone of deciduous forest to the broad plains of the Deccan with its highly variable and uncertain rainfall.

These broad ecological zones are associated with the distribution of different disease vectors. In the case of vectors for malaria, *Anopheles fluviatilis* is present down the monsoonal west coast and *A. minimus* in N.E. India and Assam and Bangladesh (Boyd, 1949, p. 811). Both these species are domestic species with marked preference for human blood and the infectivity rates for them are very high. In the past, malaria was hyperendemic in these areas, and is still not uncommon, with *P. falciparum* the chief parasite involved. In the drier areas of the plains the principal vector is *A. culifacies* and the malaria was seasonal and also epidemic, although localised areas of hyperendemicity were present due to water-logging from improperly controlled irrigation systems.

Major epidemics of malaria occurred on the plains with a periodicity of about 8 years, generally following heavy monsoonal rains after several years of unusually dry conditions. During these epidemics mortality was up to thirty times higher than normal and there was huge mortality among young children (quoted in Boyd, 1949, p. 812). Thus the selection pressure, although intense at times, was qualitatively different from that operating in the monsoon forest areas of the west and northeast.

New Guinea

New Guinea, the largest island of the Western Pacific, is a complex of ecological zones. There is a long backbone of high and rugged mountains. This highlands area provides a series of relatively isolated foci of population making up approximately 50% of the total population of New Guinea. By contrast the remaining 50% occupy the coastal areas with varying qualities of habitat from coastal mangrove and extensive swamp areas to monsoonal forest. The coastal areas themselves differ in the seasonality and extent of rainfall, so that some areas on the south coast are relatively dry. Port Moresby, the capital of Papua New Guinea, is in such an area.

The anopheline vectors of malaria in New Guinea are all species belonging to the Punctulatus complex belonging to the sub-genus Cellia. In coastal areas malaria was and, in some places, still is hyperendemic. In other places it is more seasonal, as is indicated in Figure 1 for Port Moresby. The predominant parasite during most of the year is *P. falciparum* and the peak incidence follows the very wet period at the beginning of the year. In the highland areas malaria is spasmodically epidemic and of low prevalence. How long it has been present in highland areas is a matter of argument. Some believe it is a relatively recent introduction, others that it has long history there (Ewers & Jeffrey, 1971).

In a more refined analysis of habitat and population Terrell (1976, 1977) recognised four geographical systems on Bougainville: North Coastal, South Coastal, Buin Plain and Steeplands. As in other parts of Melanesia, languages belonging to two major linguistic divisions are spoken (Austronesian and non-Austronesian or Papuan), and several attempts have been made to distinguish anthropometrically or anthroposcopically between speakers of these two types of languages. Oliver and Howells (1951), for example, concluded that on Bougainville there were three "distinctive" types of physique. these are: small Southern inland Papuan-speakers from the Buin plain: large Papuan-speakers from the Steeplands; and large, but distinctive, Austronesian-speakers from the coastal areas. However, on the basis of blood genetic marker data Terrell concluded that no one had demonstrated that Papuan-speakers as a group are biologically distinctive from Austronesian-speakers as a group, and he argued that the variability in the Bougainville populations could be understood only in terms of their biogeographical relationships. Bougainville is a valuable illustration that on even a small island (250 km long and 65 km wide) there is anything but a uniform environment. The various geographical regions have developed social patterns of interaction influenced by terrain and economic resources, and these, in turn, must inevitably have interacted with the biological characteristics of the people in these regions.

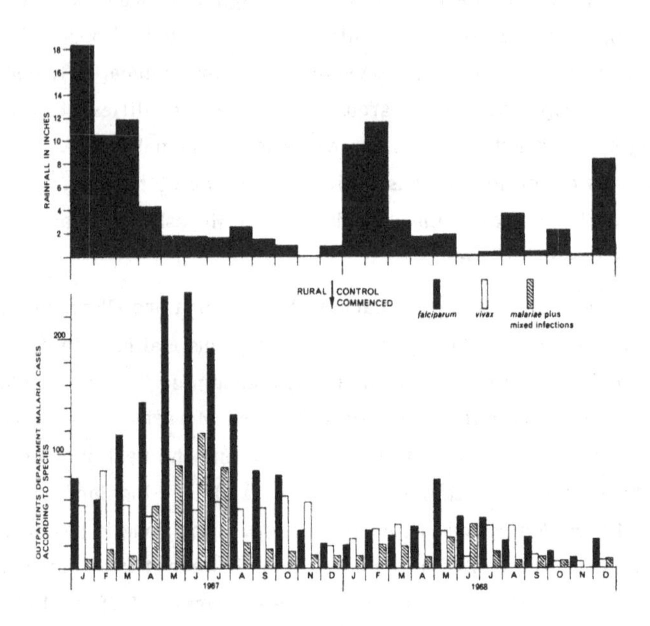

Figure 1: Seasonal variation in Malaria incidence in Port Moresby, Papua New Guinea, before and after spray control of mosquitos. (From Ewers & Jeffrey, 1971.)

GENETIC VARIABILITY: ECOLOGICALLY 'SENSITIVE' MARKERS

We have been fortunate to have studied populations across the entire tropical zone of India, S.E. Asia and the western Pacific. Our own data have been supplemented by the work of several other groups, particularly that of D. F. Roberts and his collaborators in India, of Lie-Injo Eng and her co-workers in S.E. Asia, and Keiichi Omoto and collaborators in the Philippines. These combined data cover primarily blood genetic markers: red cell and white cell surface antigens (blood groups and HLA), serum proteins, and red cell enzymes. Some additional, but limited, genetic information will be referred to below.

Among these systems there are certain genetic markers which clearly are selected by environmental agents, of which malaria is the most important. These markers I will refer to as ecologically 'sensitive': they include the abnormal haemoglobins, the thalassaemias, the G6PD deficiencies and a red-cell morphological variant, ovalocytosis.

(a) Haemoglobinopathies:

The distribution of abnormal haemoglobins and the thalassaemias has been determined in some detail for the areas under review. For the former, HbS, HbE and HbJ Tongariki achieve significant frequencies over defined parts of the range. The thalassaemias constitute an important problem both in these areas and in some places where abnormal haemoglobins are absent.

Following the detection of sickle cell disease in an Indian living in South Africa (Berk & Bull, 1943) it was subsequently shown to be present at high frequency among some of the tribal populations of South India by Lehmann and Cutbush (1952). This work was confirmed and extended by me in 1960 and by many others studying either the sickling phenomenon or identifying HbS by electrophoresis. The results were summarised by Saha and Bannerjee (1973) and more recently by Brittenham (1981). The highest frequencies of the sickle cell gene (up to 35%) occur in the hills of central and southern India among tribal populations. It is found only rarely in non-tribal Hindu or Muslim populations. There are some exceptions to this. For example, Danukhs in Uttar Pradesh have a frequency of the sickle cell trait of approximately 10%, Sorathis in Gujarat of 27%, and Mahars at Jugchalpur in Madhya Pradesh have a frequency of 38% (Brittenham, 1981). It is possible that Danukhs, Sorathis and Mahars have tribal origins, or were strongly influenced by tribal neighbours. A small number of HbS carriers have been reported among tea plantation workers in Assam, but these are people who have migrated from HbS areas in central India.

By contrast, in Assam and further east into South-East Asia and down through the Malaysian peninsula into Indonesia the predominant abnormal haemoglobin is HbE (Wasi, 1983). Indeed in some areas the gene frequency of HbE exceeds 50%. HbE also is the only abnormal haemoglobin in Sri Lanka, where it achieves quite high frequencies among the Veddahs, but is rare amongst the Sinhalese. Beyond Indonesia HbE disappears. However, in parts of the western Pacific its place is taken by an α-chain variant HbJ Tongariki. We identified this first during a study on the island of Tongariki, in Vanuatu and, in this small population, it had a frequency of 10% (Gajdusek et al, 1967). At first we thought this was a local mutation, restricted probably to this one population. However, subsequent surveys have shown it to be more widely distributed, being found in parts of New Guinea, the Solomon Islands and the Banks and Torres islands.

The distribution of these high frequency abnormal haemoglobins is, in general, coincident with areas where *falciparum* malaria is hyperendemic - HbS in west and central India, HbE in north-east India and S.E. Asia and HbJ Tongariki in the Pacific. For the first two of these there is now good evidence that they are subject to selection in a highly malarious environment. Similar evidence for HbJ Tongariki so far has not been produced, but its pattern of distribution suggests that it may be subject to selection. But why three abnormal haemoglobins? Once again, the pattern of distribution supplies some clues. HbS belongs to the pre-Dravidian or Australoid populations in India, now confined to relatively small areas. HbE is associated with those populations with a Mongoloid constellation of genes, and HbJ Tongariki is associated with speakers of Austronesian languages in the western Pacific. The fact that we do not find mixed populations of abnormal haemoglobin genes, except in areas of contact between the two, argues strongly first that the chance of a new mutant surviving and advancing in the population, even under intense malarial pressure, is not high and, secondly, that once an effective protective gene is established the chance of a new mutation entering the population is reduced.

Assessing these postulates is likely to be difficult unless the present battle against malaria is irrevocably lost. Livingstone (1983) reviewed studies over the last thirty years and concluded that the recency of malaria as a selective factor seems to have created the diversity of abnormal haemoglobins and to be the cause of nonequilibrium in most populations. He concluded (p. 31) "If malaria selection continued in human populations, much of this variation would decrease as alleles with higher fitnesses and thus a competitive advantage replaced the others. Many of these polymorphisms would thus qualify as 'fugitive' alleles. If this view of the malaria hypothesis is accepted, then the most important task now facing the analysis of red cell variation is the estimation of the relative fitness of the genotypes. At present, this seems to be an impossible task and it may be the only criterion to judge different estimates will be their plausibility in explaining this variation".

(b) *The thalassaemias*

The difficulty of screening for thalassaemia traits makes our knowledge of the distribution of these conditions much less secure than for the abnormal haemoglobins. However, Wasi (1983) suggested there are three distinct zones in Asia.

(i) North Asia where the thalassaemias are absent,

(ii) South-east Asia where both α- and β-thalssaemia are present, and

(iii) North India into the Middle East where the thalassaemias are predominantly of the β-type.

In the past, the frequency of the α-thalassaemias were most readily estimated from the frequency of the Hb Bart's in the cord blood of newborn infants, or of HbH in adult blood. In India Hb Barts has been reported from West Bengal and Maharashtra and presumably occurs in populations in between. Hb H has been reported sporadically from across the north of the country. Little is known of the type of α-thalassaemia, either in the north or the south. Indirect evidence for α-thal/HbS in some tribal populations in the south suggests it is not uncommon in these tribal populations (Brittenham et al, 1977, 1979). Other studies by Brittenham et al (1980) among three tribal groups in Andhra Pradesh found a high proportion with one α-gene deletion, but not of chromosomes with two α-gene deletions.

In S.E. Asia, α-thal, as judged by the frequency of Hb Barts in cord blood, is relatively widespread and common. Further to the south-east, New Guinea also has a high frequency of α-globin gene deletion as defined recently by Weatherall and his colleagues and by Yenchitsomanus and his collaborators in our laboratory (Yenchitsomanus et al, 1985). The latter group has found extremely high frequencies of the deletion form of α-thalassaemia (-α/), as studied by DNA mapping in Madang on the north coast of Papua New Guinea. 97% of the population tested from Madang and 89% from nearby KarKar island were either -α/ heterozygotes or homozygotes. By contrast, no examples of the deletion form were detected in the Eastern Highlands of PNG. Of the deletions 96% were of the 4.2 kb type, with only 4% of the 3.7 Kb type - the only population reported so far where the 4.2 Kb type is the predominant form.

Yenchitsomanus et al (1985) conclude that, "the presence in high frequencies of α^+-thalassaemia in the coastal area of Madang and on the neighbouring island, where malaria has long been holoendemic or hyperendemic, and its virtual absence from the non-malarious highlands of PNG, suggest the role of malaria as the selective factor in maintaining α^+-thalassaemia. If this selective pressure is still operating and since α^+-thalassaemia has no apparent homozygous disadvantage, the abnormal haplotype (-α/) will be in the process of fixation in this population".

By contrast with the α-thalassaemia, the β-thalassaemias have been studied more intensively in the region under discussion. In India Chatterjea

(1966) at the School of Tropical Medicine in Calcutta, reviewed 796 cases, of which 193 were apparent β-thalassaemia cases and the majority of the remainder were compound heterozygotes mostly HbE/Thal, with some cases of HbD/thal. He also reported 157 homozygous β-thal cases from Bombay, and homozygous β-thal has now been reported from many areas in northern India, with heterozygote frequencies ranging from 6.5% in West Bengal, 7% in Madhya Pradesh, 1.6-5.5% among various tribals in Assam and from 9.2 to 17% among various groups of Lohanas in Maharashtra. Figures for south India suggest a generally low prevalence of β-thal, although cases have been reported from each of the four States in the South and a study from Pondicherry has reported a high 16.2% in 142 persons studied. δβ-thalassaemia also appears to be widespread in India, with most cases reported coming from the north.

In S.E. Asia, Wasi (1983) pointed out that although cases of β-thalassaemic disease are documented from all parts, especially cases of HbE/thal, there are relatively few in terms of frequency, varying between 1 and 9%. Flatz and his colleagues (1965) first pointed out a reciprocal relationship between HbE and β-thalassaemia. The explanation for this may be that HbE/βthal leads to a disease almost as severe as homozygous β-thal, resulting in gene loss. But HbE has a higher fitness, homozygous E/E persons being normally productive whereas homozygous β-thal has almost zero fertility. Where both genes coexist, therefore, E will spread at the expense of β-thal. β-thal also is widespread in lowland areas of New Guinea, but little information is available for other parts of the western Pacific. Further information is still inadequate for the molecular basis of the β-thalassaemias in the whole region. Studies in progress at the present time should improve this situation in the near future, but much additional work still remains to be done.

(c) *Deficiency of glucose-6-phosphate dehydrogenase:*

Information on the distribution of G6PD deficiencies is less reliable than for the haemoglobinopathies. This is due in part to the unreliability of some of the technical methods employed and also to the instability of the enzyme which leads to false positives for deficiency in specimens which have been improperly handled. However, the general picture for the region under consideration is similar to that for the thalassaemias. In India there is a widespread level of deficiency at frequencies of 1-5% for males, with some reports of much higher levels in certain groups, for example, 10% among the Mahars in one report and 17% among Parsis (Livingstone, 1967).

Frequencies in S.E. Asia also are variable, with relatively high frequencies in Thailand (10-15%) and values of 1-6% in Malaysia and Indonesia, although they are 12% for Malays in Sarawak and 24% for the Murut in Sabah. New Guinea again shows extreme values, high in most coastal areas and very low in the highlands. In some lowland populations frequencies of 30% are found, whilst most populations in the highlands have no deficiency or 1-2%. Variable frequencies have been reported for other western Pacific Islands, with Polynesian populations in general having no deficiency (Livingstone, 1967).

Where electrophoretic studies have been carried out the deficiency is not of the African A-type. What is more important is the evidence for multiple mutations and selection in a malarious environment. Chockkalingam and collaborators, in our department, recently have purified and characterised biochemically the enzyme defect in a survey of persons from various parts of Papua New Guinea (Chockkalingham et al, 1982). In a screen of 362 males they identified 26 deficient individuals. After purification of the enzyme they found 13 new variants in addition to copies of previously described variants - bringing the total of variants in New Guinea to more than 20. Some of the variants (Gd Markham and Gd Salata) are widespread, whilst others (Gd Goodenough or Kaluan) are restricted to one geographic locality. This suggests that some mutants are older and were carried to different places in New Guinea by the early Austronesian invaders. Others are more recent mutations which have remained localised and have increased in frequency under selection pressure.

On the basis of the heterogeneity demonstrated in these New Guinea studies it will be of interest to characterize more fully the G6PD variants from various parts of S.E. Asia and India, as well as from the other islands of the western Pacific.

(d) *Ovalocytosis:*

A morphological variant of human red cells, variously called elliptocytosis or ovalocytosis, has been known in S.E. Asian populations for nearly 50 years. Bonne and Sandground (1939) reported this first for the Toradja living around Lake Lindu in Central Sulawesi with a frequency of nearly 50%. Subsequently Lie-Injo and her colleagues reported very high frequencies of ovalocytosis among aboriginal Malays, although the frequency in Malays themselves is low (Lie-Injo, 1976). However, detailed studies in Papua New Guinea by Dr. Serjeantson together with others, have shown the possible relationship between this trait and resistance to malaria. Serjeantson et al (1977) pointed out the restriction of ovalocytosis to areas of endemic malaria. Further they studied the severity of infection in a sample of ovalocytic children compared with normocytic children. The ratio of parasitaemia for ovalocytosis compared to normocytics was 1.05 for *P. falciparum*, 0.90 for *P. vivax* and 0.54 for *P. malariae* and 0.91 for infection with any species.

Kidson et al, (1981) confirmed the effect of ovalocytosis on parasitisation, making use of cultured *P. falciparum* tagged with fluorescent antibody. They showed that ovalocytic red cells from Papua New Guineans are very resistant to infection by *P. falciparum* in culture, using a double label fluorescent assay of merozoite invasion.

In Papua New Guinea, therefore, where abnormal haemoglobins do not occur, there are three types of ecologically sensitive genes involved in resistance to malaria: thalassaemias, G6PD deficiencies and the gene for ovalocytosis. The dynamics of the interaction between these genes has not been worked out. But in recent unpublished work from our department Sofro has shown an inverse relationship in Indonesia between the frequency of ovalocytosis and G6PD deficiency (Sofro, 1982). This is similar to the relationship demonstrated in Thailand by Flatz et al (1965) between HbE and β-thalassaemia.

GENETIC VARIABILITY: OTHER GENETIC MARKERS

In addition to the ecologically 'sensitive' markers data have been published for genes at a large number of loci controlling red and white cell antigens, serum protein groups and red cell enzymes. Data on the red cell antigens have been definitively collated by Mourant and his colleagues (1976), and further, it is not possible in the space available to review adequately all the salient features of the variation in the non-red cell antigen systems for the region of interest to us now. I intend only to draw attention to some of the more informative aspects of this variation, and I will do this under two headings: specific, geographically restricted alleles, and variation in commonly polymorphic systems. In some cases, of course, geographically restricted alleles occur in systems which also are universally polymorphic and, in some cases, recent development in iso-electric focussing have placed systems in the first category into such a compound category.

(a) *Specific, geographically restricted alleles:*

(i) Lactate dehydrogenase:

Lactate dehydrogenase (LDH) is monomorphic in most human

populations, and variants are rare. We have found two exceptions to this, one among populations in India, the other restricted to one group in the Western Highlands of Papua New Guinea. In the latter locality there were 6 persons with an LDH A subunit variant among 406 members of the Murapin clan of the Enga. We have found no other examples of this particular variant among many thousands of New Guineans tested (Blake et al, 1969).

In India the situation is markedly different. Our first report found an LDH variant with a frequency of 1.6% among Bengalis in Calcutta. This Calcutta-1 variant has unusual properties for the distribution of its isozymes, but we have now detected it in many parts of India, at frequencies ranging from zero to nearly 4%. We have identified also two persons homozygous for Cal-1, one in south India and the other in Bombay (Das et al, 1970; Saha et al, 1974).

(ii) Malate dehydrogenase:

Malate dehydrogenase (MDH), at least the cytosol form, also is monomorphic in most human populations, with a number of rare variants being reported in various populations (review in Blake 1978). In New Guinea, however, we first reported an S-MDH variant, S-MDH3 at polymorphic frequency (Blake et al, 1970). Frequencies of the variant allele in New Guinea vary from zero to 7%. The lowest values occur, on average, among populations in the eastern coastal areas, almost all predominantly speakers of Austronesian languages. The higher frequencies occur among Papuan-speaking peoples, who still occupy the greater part of Papua New Guinea and Irian Jaya. Other examples of this variant have been detected by Sofro in Macassarese and Toraja in Indonesia (Sofro, 1982).

Another example of a polymorphic S-MDH variant was detected by us among Micronesians in the Western Caroline Islands. Here, 5% of more than 700 persons tested had an S-MDH6 variant. Work by Blake in our laboratory revealed interesting characteristics which suggested it had a reactive sulphydryl group. However, so far this has not been substantiated (Blake, 1978).

(iii) Carbonic anhydrase:

Variants at both the CA_1 or CA_2 loci ("low" and "high" activity forms respectively) are rare, except for variants at the CA_2 locus, which is polymorphic among Black Africans. A polymorphic CA_1 variant was reported first by Tashian et al (1963) in Guam. Since then identical variants have been found in various parts of the S.E. Asian region with the highest frequency occurring among the Mamanwa negritos in the Philippines (Omoto et al, 1981). The Philippines, therefore appear to be the focus for this allele, which has been dispersed to other parts of the Pacific and S.E. Asia from there.

We have, of course, more recently reported other polymorphic alleles at both the CA_1 and CA_2 loci among Australian Aborigines, but that is outside the scope of the present survey.

(iv) Thyroxine-binding globulin:

A recent discovery by Kamboh, in our laboratory, has thrown new light on the distribution of variants at the TBG locus. Previously, only Black Africans (and Black Americans) have been found to be polymorphic for TBG. Kamboh and Kirwood (1984) noted a slow TBG variant, indistinguishable electrophoretically from the African variant, which was found in 1-10% of all the Melanesian and Polynesian series studied and at low frequency in some Micronesian and Indonesian series. However, it has not been detected in any Indian population or in Australian Aborigines.

(b) *Universal polymorphic systems:*

Discussion is restricted to those which also have additional alleles which are geographically localized, and are therefore informative about the relatedness or otherwise of populations which are of interest to us here.

(i) Transferrin:

Kamboh and Kirk (1983) surveyed the distribution of transferrin alleles as revealed by isoelectric focussing. Two subtypes of the common TfC allele are found in all populations studied (TfC1 and TfC2) whereas TfC3, which is relatively common in Europeans, achieves appreciable frequencies only in north Indians, with somewhat reduced frequency in South India and it is absent in tribal populations sampled in South India. TfC3 appears to be a European gene marker brought into India by population movement and hybridization.

In addition to the subtypes of C, other specific alleles at the Tf locus are of importance. Nearly twenty years ago I drew attention to the high frequencies of TfD1 in Melanesians and Australian Aborigines (Kirk, 1968). By contrast, another allele, TfDChi, with very similar electrophoretic mobility is present in East Asian populations and, incidentally, also in many Amerindian populations. Dr. Salam Sofro (unpublished) has found that Indonesia, particularly the Lesser Sunda Islands, is a hybrid zone where both D1 and DCh1 are present (Sofro, 1982).

(ii) Vitamin D-binding protein:

More than 20 years ago we drew attention to the importance of specific alleles at the vitamin D-binding, or Group Specific (Gc) locus (Cleve et al, 1963), reporting at that time a new allele, Gc Aborigine. This is widely distributed among Melanesian populations in the Western Pacific and in Australian Aborigines. It is indistinguishable by present methods from a similar allele in Black Africans. It is also present occasionally in the Lesser Sunda Islands of Indonesia and among Polynesians and in Nauru (Kamboh et al, 1984).

Using isoelectric focussing we have studied also the distribution of the subtypes of Gc1 (Kamboh et al, 1984). The highest frequency of the 1S allele (>50%) is found among Indians, reflecting genetic affinity with Europeans. The Gc2 allele has its highest frequency among Polynesians (with some groups up to 40%) whilst the lowest is among Australian Aborigines, with less than 10%.

(iii) PGM locus 1 and 2:

Two PGM loci are expressed in human red cells. PGM_1 is universally polymorphic, whilst PGM_2 is polymorphic in only a limited number of populations. The first polymorphic allele at locus 2 to be described was PGM_2^2, which is present with frequencies up to 5% in Black Africans. Other alleles have been reported by Santachiara-Benerecetti in African and Indian populations, but none at polymorphic frequency. However, several PGM_2 alleles are polymorphic in the Western Pacific area. PGM_2^9 achieves its highest frequencies in New Guinea (14%) among the Dani in West Irian and among the Anga and Chimbu in the Eastern Highlands of Papua New Guinea. This allele is not restricted to non-Austronesian speakers: it has a frequency of 2% among the Motu-speakers in Port Moresby and is found also in some populations of the Banks and Torres Islands and even in the Polynesian outliers of Anuta and Tikopia (Blake & Omoto, 1975).

Another PGM_2 allele, PGM_2^{10} also is widely distributed in the Western Pacific. It reaches frequencies of almost 10% in parts of West Irian and is present in the Eastern Highlands as well as in some of the Torres Islands and occasionally elsewhere. A single case has been reported in an Australian Aborigine from Arnhem Land (Blake & Omoto, 1975).

PGM_2^9 has also been found widely in Indonesian populations, but only in one (outside West Irian) does it reach polymorphic frequency, amongst rural Javanese. By contrast, PGM_2^{10} is rare in Indonesia, the only example outside West Irian being in a Balinese village (Sofro, 1982).

The distribution of alleles at the PGM_1 locus is much more complex. Kamboh and Kirk (1984) have reviewed this topic in detail, making use of isoelectric focussing techniques. In brief, the 1+ allotype frequency varies from a low of 43% among Polynesians in Samoa to 77% among Melanesians in Port Moresby. Melanesians and Indonesians have relatively high frequencies of 1- (21-25%), whereas east Asians and Indians have values approximately half of this. Indian populations, however, have high values of the 2+ allotype (22-31%) whereas the lowest value of 2% is recorded in Micronesians.

Several specific alleles at the PGM_1 locus are important also. The first of these to be reported from England, PGM_1^3 appears to be widespread, though definite proof of the identity of the variant PGM_1^3 alleles reported is still awaited. In the Pacific area it has been found in Japan, Malaysia and India, although not at polymorphic frequency. On the other hand, PGM_1^3 is common in many parts of New Guinea and in some places in Micronesia, the Banks and Torres Islands and in the Solomon Islands (Blake and Omoto, 1975).

Similarly, PGM_1^7 has been reported from Japan and with low frequency in SE Asia and also in India. Amongst the Micronesians, in the Western Caroline Islands, it achieves high frequencies (4% on Ulithi and 8% on Woleai). Recently we have studied examples of PGM_1^3 and PGM_1^7 using isoelectric focussing techniques and have distinguished both 3+ and 3- subtypes among New Guinean and Micronesian samples. For PGM_1^7, alleles in the Pacific area appear to be of one type only, but in India we have distinguished another type, both in the north and in the south. In Japan Takahashi (1982) has reported 7+ and 7- subtypes of

PGM_1^7. Our Indian variant appears to be identical with Takahashi's 7+ subtype, whilst our PGM_1^7 variants found at polymorphic frequency among Micronesians appear to be the same as his 7- subtype. Confirmation of these designations is awaited with interest (Kamboh & Kirk, 1984).

(iv) Gm:

The complex haplotypes of the Gm-Am systems are a valuable tool for tracing population movements and for demonstrating variation within populations. Two areas within the region illustrate the power of the Gm-Am system. In India, Schanfield and Kirk (1981) found a marked north to south clinal variation in the haplotype distribution, with $Gm^{z,a:g}$ varying from 23 to 38% and $Gm^{z,a,x:g}$ from 7 to 18%. Inverse variation was observed for $Gm^{f:b}$ (from 50 down to 30%) and $Gm^{z,a:b}$ (15 down to 5%). These differences are further magnified between high castes in the north and low castes in the south. In addition, high castes in both the north and the south have indications of Asian admixture. We believe that the $Gm^{z,a:b}$ haplotype found in India is of central Asian origin. The frequency of $A_2m(2)$ positive specimens shows a tendency to increase from north to south and, again, this difference is accentuated between high castes in the north and low castes in the south.

In the Western Pacific, the most detailed studies of Gm-Am have been concentrated in New Guinea and Australia (summarised by Schanfield, 1977). Among speakers of non-Austronesian languages the allotype $Gm^{za:n:b}$ combined with $A_2m(1)$ is common, whereas for speakers of Austronesian languages $Gm^{fa:n:b}$ combined with either $A_2m(1)$ or $A_2m(2)$ is common. However, in some parts of New Guinea these two sets of allotypes are not clear-cut markers of language differences.

In Australia there is also a dichotomy between the central and northern parts of the continent. In the centre only $Gm^{za:n-:g}$ and $Gm^{zax:n-:g}$ are present, mainly in combination with $A_2m(1)$. In the north, however, $Gm^{za:n:b}$ with $A_2m(1)$ is present also. The latter allotype represents, probably, relationships between northern Australia and New Guinea, when a second wave of non-Austronesian speakers entered the area. Clearly, far more detailed information is needed about the distribution of the Gm and A_2m groups in the whole of the region.

(v) HLA:

Almost certainly, the most informative of all systems at the present time is the human leucocyte antigen system and the associated complement loci on the short arm of chromosome 6. Already, recombinant DNA techniques are revolutionising the study of HLA and complement types and within a few years considerably more detailed information will be available to assist in understanding population structure and human migrations.

Summarising what is known about this at present, extensive data are available only for the HLA A and HLA B loci. India is still little studied, though what information we have indicates that northern India in particular has some European features, but obviously more detailed information is needed from several parts of the country (Mittal, 1982).

For the Pacific area, Dr. Serjeantson has emphasized the power of HLA for discriminating between populations on the basis of HLA A and B data alone. In a recent publication (Serjeantson, 1985) she has shown that populations cluster together on the basis of ethnic groups (Polynesian, for example, in one cluster) and hybrid populations also demonstrate their position relative to the parent populations. She has shown also that linkage disequilibrium values between HLA A and B alleles are valuable in demonstrating ancestral relationships. Australian Aborigines do not cluster with any other groups, but linkage relationships clearly indicate the persistence of Australoid elements in Melanesia. She has also given maximum likelihood estimates of admixture based on HLA A and B frequencies for Fiji, which suggests that the present Fijian population is a mixture of about 20% Polynesian, about 20% Austronesian, and 60% non-Austronesian. A similar calculation for Guam shows that present-day Guamanians are 36% European, 17% Filipino and the remainder Chamorros.

GENETIC VARIABILITY: MULTIVARIATE ANALYSES

The maintenance of genetic variability is important for the potential for selection in the future. Reduction in genetic variability over the whole genome therefore is not an advantage. However in populations this can be brought about by inbreeding, or by passing through bottlenecks which reduces the breeding population to a very small number of individuals. Some of the human populations in the region we are now considering practise inbreeding and others have experienced bottlenecks.

Inbreeding is a common feature of Indian populations, particularly Muslim populations in the north and both Hindu and Muslim populations in the south. Sundar Rao (1984) recently has summarised both the extent of inbreeding in India, and the effects of inbreeding on biological variables. Making use of genetic marker data for populations in northern India, Papiha et al (1982) have calculated Wright's F statistics to determine the extent of genetic variability within and between populations. Their F_{ST} values (that is the coefficient of genetic differentiation among sub-populations) show considerable genetic differentiation among the populations, which they believe reflects the ethnic diversity of the area they were studying. However local variation within each State is lower, indicating a geographical component to the total variation.

Another simple way of looking at variability is to calculate the heterozygosity. The absolute heterozygosity, of course, depends on the loci being considered and also on the technical methods used for revealing allelic variation. Recently, Dr. Kamboh, in our department examined a series of populations in India, S.E. Asia and the Pacific, comparing the heterozygosity values obtained using starch gel or agarose gel techniques with those found using isoelectric focussing. In each of the four systems the heterozygosity was increased, dramatically for Pi and Tf, and was almost doubled for Gc and PGM_1.

Table 1 shows that the mean heterozygosity based on these four systems is relatively high in all the populations studied. The highest values (above 0.5) are in the continental populations. With the exception of the Lesser Sunda Islands and Fiji the Pacific islanders have lower heterozygosities, with Micronesians in Nauru and Kiribati having the lowest values (0.413 and 0.401 respectively). This suggests that many Pacific Island populations have passed through bottlenecks which have reduced their genetic variability.

Finally, the genetic variability can be used to explore the evolutionary relationships between populations in the region. Clearly the greater the variability, as measured by the heterozygosity, the better the discrimination will be. Using Nei's distance statistic, Dr. Kamboh has produced dendrograms and eigenvector diagrams making use of the large amount of information for the four loci just discussed (Pi, Gc, Tf and PGM_1) using the data derived from IEF studies (Figures 2 and 3). For the dendrogram, the most extreme differentiation is one branch with New Guinea Highlanders and Australian Aborigines, with another

Table 1: Heterozygosities for various populations for PGM1, Tf, P1 and Gc

Population	IEF PGM	IEF Tf	IEF P1	IEF Gc	Mean (all systems)
Thailand	0.567	0.430	0.453	0.640	0.523
India:					
Delhi	0.600	0.409	0.564	0.578	0.538
Madras	0.640	0.366	0.534	0.593	0.533
Soliga (tribal)	0.653	0.451	0.601	0.413	0.530
Indonesia:					
Lesser Sunda Is.	0.578	0.316	0.565	0.614	0.518
Cook Islands	0.537	0.355	0.380	0.631	0.476
American Samoa	0.657	0.170	0.332	0.665	0.456
Western Samoa	0.703	0.189	0.373	0.663	0.482
Wallis Island	0.625	0.320	0.272	0.662	0.470
Kiribati	0.478	0.090	0.405	0.630	0.401
Nauru	0.396	0.086	0.510	0.658	0.413
New Caledonia	0.606	0.137	0.434	0.643	0.455
Fiji	0.695	0.270	0.507	0.680	0.538
Papua New Guinea:					
Port Moresby	0.377	0.214	0.505	0.673	0.442
Sepik River	0.544	0.283	0.530	0.690	0.512
Eastern Highlands	0.464	0.325	0.465	0.738	0.498
Central Australia	0.453	0.183	0.482	0.645	0.441
Mean (all populations)	0.556	0.282	0.457	0.632	0.482

distinct branch with African (both those in Africa and in America), whilst a third branch has Europeans and Asiatic Indians. The remaining Asian and Pacific populations fall into a final loose cluster. Chinese, Japanese and Thais form one tight grouping and one Polynesian population is included in this grouping. Two other Polynesian populations, Samoans and Wallis Islanders, are a little more separate. Another sub-grouping links the Lesser Sunda Islands with Fijians, whilst a final sub-group includes Micronesians from Nauru and Kiribati together with New Caledonians.

The inclusion of sub-typing in any one of the four systems considered here makes possible discrimination between the populations broadly in accord with accepted views on the relationship between peoples in Asia, Europe and Africa. Combining all four systems together enhances this discriminatory power and provides a pattern of relationships similar to that given by Serjeantson (1985) using data for HLA. In both the previous studies of Kirk (1980) and of Serjeantson and her colleagues, as well as those referred to above, Polynesians

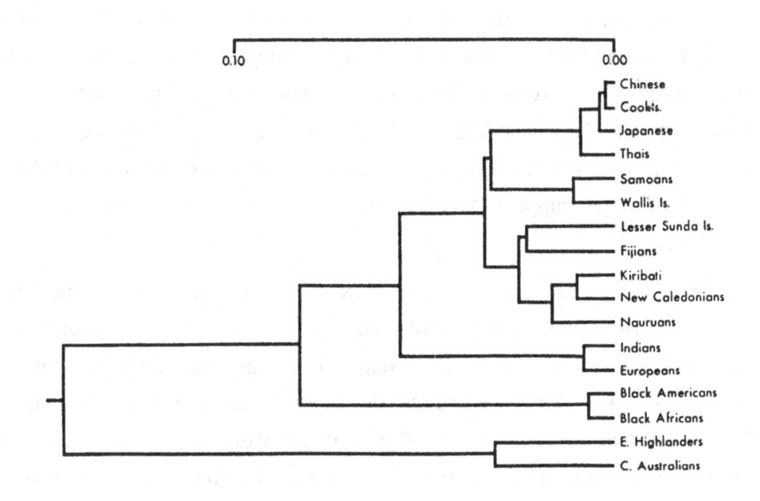

Figure 2: Dendrogram showing population relationships based on IEF data for Pi, Tf, Gc and PGM_1. (From Kamboh, 1984.)

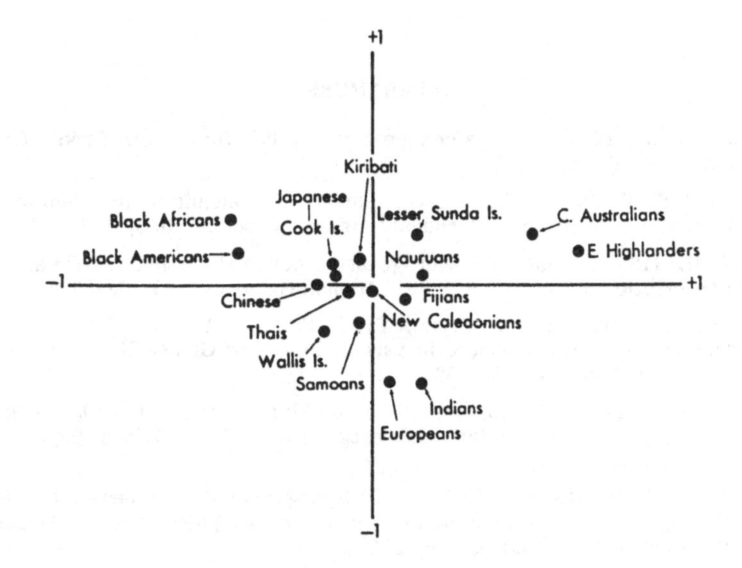

Figure 3: Eigenvector diagram for the same data used in Figure 2. (From Kamboh, 1984.)

are close to east Asians in their genetic affinities. This is in agreement with the view espoused most strongly by Bellwood (1979).

By contrast, the two Micronesian populations included in our present study, Kiribati and Nauru, are not clearly distinguished from the population of New Caledonia. Other evidence from specific genetic markers, such as complement C6 Nauru, indicates considerable gene flow between Nauru and other western Pacific populations, and the close relationship found in the present investigation adds support to the extent of the gene flow between these populations.

Finally, the extreme position of New Guinea Highlanders and Australian Aborigines is reinforced by the subtyping results. Populations moving into New Guinea and Australia have a time depth much greater than elsewhere in the Pacific, except for Indonesia and the Philippines which, in the Pleistocene, were part of the Asian continent. But this greater time depth has permitted differentiation from later populations moving into the area. It is now clear that, although New Guinea populations are genetically quite distinct from Aboriginal Australians, together they are quite distinct from any of the other populations in the Pacific.

REFERENCES

Bellwood, P. S. (1979). Man's conquest of the Pacific. New York: Oxford University Press.

Berk, L. & Bull, G. M. (1943). Case of sickle cell anaemia in an Indian woman. Journal of Capetown Postgraduate Medical Association, **2**, 147-152.

Blake, N. M. (1978). Malate dehydrogenase types in the Asian-Pacific area and a description of new phenotypes. Human Genetics, **43**, 69-80.

Blake, N. M., Kirk, R. L., Pryke, E. & Sinnett, P. (1969). Lactate dehydrogenase electrophoretic variant in a New Guinea Highland population. Science, **163**, 701-702.

Blake, N. M., Kirk, R. L., Simons, M. J. & Alpers, M. P. (1970). Genetic variants of soluble Malate dehydrogenase in New Guinea populations. Humangenetik, **11**, 72-74.

Blake, N. M. & Omoto, K. (1975). Phosphoglucomutase types in the Asia-Pacific area: a critical review including new phenotypes. Annals of Human Genetics, London, **38**, 251-273.

Bonne, C. & Sandground, J. H. (1939). Quoted in Lie-Injo (1976).

Bowles, G. T. (1977). The People of Asia. London: Weidenfeld and Nicolson.

Boyd, M. F. (ed) (1949). Malariology Vol. 2., p. 812. Philadelphia and London: W. B. Saunders.

Brittenham, G. M. (1980). Thalassemia in southern India. Acta Haematologica, **63**, 43-48.

Brittenham, G. M. (1981). The geographic and ethnographic distribution of hemoglobinopathies in India. In: J.E. Bowman (ed.), Distribution and Evolution of Hemoglobin and Globin Loci, pp. 169-178. New York: Elsevier.

Brittenham, G. M., Lozoff, B., Harris, J. W., Sharma, V. S. & Narasimhan, S. (1977). Sickle cell anemia and trait in a population of southern India. American Journal of Hematology, **2**, 25-32.

Brittenham, G. M., Lozoff, B., Harris, J. W., Mayson, S. M., Miller, A. & Huisman, T. H. J. (1979). Sickle cell anemia and trait in southern India: further studies. American Journal of Hematology, **6**, 107-123.

Chatterjea, J. B. (1966). Haemoglobinopathies, glucose-6-phosphate dehydrogenase deficiency and allied problems in the Indian subcontinent. Bulletin of the World Health Organization, **35**, 837-856.

Chockkalingham, K., Board, P. G. & Nurse, G. T. (1982). Glucose-6-phosphate dehydrogenase deficiency in Papua New Guinea: the description of 13 new variants. Human Genetics, **60**, 189-192.

Cleve, H., Kirk, R. L., Parker, W. C., Bearn, A. G., Schacht, L. E., Kleinman, H. & Horsfall, W. R. (1963). Two genetic variants of the group-specific component of human serum: Gc Chippewa and Gc Aborigine. American Journal of Human Genetics, **15**, 368-379.

Das, S. R., Mukherjee, B. N., Das, S. K., Ananthakrishnan, R., Blake, N. M. & Kirk, R. L. (1970). LDH Variants in India. Humangenetik, **9**, 107-109.

Ewers, W. H. & Jeffrey, W. T. (1971). Parasites of Man in Niugini. Brisbane: Jacaranda Press.

Flatz, G., Pik, C. & Sringam, S. (1965). Haemoglobin E and β-thalassaemia: their distribution in Thailand. Annals of Human Genetics, **29**, 151-170.

Gajdusek, D. C., Guiart, J., Kirk, R. L., Carrell, R. W., Irvine, D., Kynock, P. A. M. & Lehman, H. (1967). Haemoglobin J. Tongariki (α115 alanine $\rightarrow$ aspartic acid): the first new haemoglobin variant found in a Pacific (Melanesian) population. Journal of Medical Genetics, **4**, 1-6.

Kamboh, M. I. (1984). Population genetic studies using isoelectric focussing in the Asian, Pacific and Australian Area. Ph.D. thesis. Canberra: Australian National University.

Kamboh, M. I. & Kirk, R. L. (1984). Genetic studies of PGM1 subtypes: population data from the Asian-Pacific area. Annals of Human Biology, **11**, 211-219.

Kamboh, M. I. & Kirk, R. L. (1983). Distribution of transferrin (Tf) subtypes in Asian, Pacific and Australian Aboriginal populations: evidence for the existence of a new subtype Tf^{6}. Human Heredity, **33**, 237-243.

Kamboh, M. I. & Kirwood, C. (1984). Genetic polymorphism of thyroxin-binding globulin (TBG) in the Pacific area. America Journal of Human Genetics, **36**, 646-654.

Kamboh, M. I., Ranford, P. R. & Kirk, R. L. (1984). Population genetics of the vitamin D-binding protein (Gc) subtypes in the Asian-Pacific area: description of new alleles at the Gc locus. Human Genetics, **67**, 378-384.

Kidson, C., Lamont, G., Saul, A. & Nurse, G. T. (1981). Ovalocytic erythrocytes from Melanesians are resistant to invasion by malaria parasites in culture. Proceedings of the National Academy of Science, U.S.A., **78**, 5829-5832.

Kirk, R. L. (1968). The world distribution of transferrin variants and some unsolved problems. Acta Genetica Medicae Gemellologiae, **17**, 613-640.

Lehmann, H. & Cutbush, M. (1952). Sickle cell trait in southern India. British Medical Journal, **1**, 404-405.

Lie-Injo, L. E. (1976). Genetic relationships of several aboriginal groups in South East Asia. In: R.L. Kirk & A.G. Thorne (eds.), The Origin of the Australians, pp. 277-306. Canberra: Australian Institute of Aboriginal Studies.

Livingstone, F. B. (1967). Abnormal Hemoglobins in Human Populations. Chicago: Aldine.

Livingstone, F. B. (1983). The malaria hypothesis. In: J.E. Bowman (ed.), Distribution and Evolution of Hemoglobin and Globin Loci, pp. 15-35. New York: Elsevier.

Mittal, K. K., Naik, S., Sansonetti, N., Cowherd, R., Kumar, R. & Wong, D. M. (1982). The HLA antigens in Indian Hindus. Tissue Antigens, **20**, 223-226.

Mourant, A. E., Kopec, A. C., & Domaniewska-Sobczak, K. (1976). The Distribution of the Human Blood Groups and other Polymorphisms. London: Oxford University Press.

Oliver, D. L. & Howells, W. W. (1951). Microevolution: cultural elements in physical variation. American Anthropologist, **59**, 965-978.

Omoto, K., Ueda, S., Goriki, K., Takahashi, N., Misawa, S. & Pagaran, I. G. (1981). Population genetic studies of the Philippines Negitos III. Identification of the carbonic anhydrase-variant with CAI-Guam. American Journal of Human Genetics, **33**, 105-111.

Papiha, S. S., Mukherjee, B. N., Chahal, S. M. S., Malhotra, K. C., & Roberts, D. F. (1982). Genetic heterogeneity and population structure in north-west India. Annals of Human Biology, **9**, 235-251.

Rao, P. S., Sundar (1984). Inbreeding in India: concepts and consequences. In: J.R. Lukacs (ed.), The People of South Asia: the Biological Anthropology of India, Pakistan and Nepal, pp. 239-268. New York and London: Plenum Press.

Saha, N. & Bannerjee, B. (1973). Haemoglobinopathies in the Indian subcontinent. Acta Genetica Medicae Gemellologiae, **22**, 117-138.

Saha, N., Kirk, R. L. & Undevia, J. V. (1974). Two further cases of homozygosity for a lactate dehydrogenase variant gene. American Journal of Human Genetics, **26**, 723-726.

Schanfield, M. S. (1977). Population affinities of the Australian Aborigines as reflected by the genetic markers of immunoglobins. Journal of Human Evolution, **6**, 341-352.

Schanfield, M. S. & Kirk, R. L. (1981). Further studies on the immunoglobulin allotypes (Gm, Am and Km) in India. Acta Anthropogenetica, **5**, 1-21.

Serjeantson, S. W. (1985). Migration and admixture in the Pacific: insights provided by human leucocyte antigens. In: R. Kirk & E. Szathmary (eds.), Out of Asia: Peopling the Americas and the Pacific, pp. 133-146. Canberra: Journal of Pacific History.

Serjeantson, S., Bryson, K., Amato, D. & Babona, D. (1977). Malaria and hereditary ovalocytosis. Human Genetics, **37**, 161-167.

Sofro, A. S. M. (1982). Population Genetic Studies in Indonesia. Ph.D. thesis. Canberra: Australian National University.

Tashian, R. E., Plato, C. C. & Shows, J. R. (1963). Inherited variant of erythrocyte carbonic anhydrase in Micronesians from Guam and Saipan. Science, **140**, 53-54.

Takahashi, N., Neel, J. V., Satoh, C., Nishizaki, J. & Masunari, N. (1982). A phylogeny for the principal alleles of the human phosphoglucomutase-1 locus. Proceedings of the National Academy of Science, U.S.A., **79**, 6636-6640.

Terrell, J. (1976). Island biogeography and Man in Melanesia. Archives of Physical Anthropology in Oceania, **11**, 1-17.

Terrell, J. (1977). Geographic systems and human diversity in the North Solomons. World Archaeology, **9**, 72-81.

Wasi, P. (1983). Hemoglobinopathies in Southeast Asia. In: J.E. Bowman (ed.), Distribution and Evolution of Hemoglobin and Globin Loci, pp. 179-208. New York: Elsevier.

Yenchitsomanus, P-T, Summers, K. M., Bhatia, K. K., Cattani, J. & Board, P. G. (1985). Extremely high frequencies of α-globin gene deletion in Madang and on Kar Kar Island, Papua New Guinea. American Journal of Human Genetics, **37**, 778-784.

MALARIA PROTECTIVE ALLELES IN SOUTHERN AFRICA: RELICT ALLELES OF NO HEALTH SIGNIFICANCE?

T. JENKINS and M. RAMSAY

MRC Human Ecogenetics Research Unit,
South African Institute for Medical Research,
University of the Witwatersrand, Johannesburg

INTRODUCTION

The history of the evolution of malaria is not known with any degree of certainty. It is found among all the major groups of terrestrial vertebrates in the Old World, suggesting that the parasites predate the hominids by millions of years. It has been claimed that plasmodial species evolved in tropical Africa or else in South-East Asia. Bruce-Chwatt (1965) pointed out that there were references to malaria dating back to 2000 BC and Hippocrates writing in the 5th Century BC was well aware of malarial fevers. There are also references to malarial fevers in the literature of India, and one writer went so far as to attribute them to mosquitos.

Prior to the 3rd Century BC, the entire population of sub-Saharan Africa consisted of small bands of hunter-gatherers. Such a low population density, coupled with the fact that such people were not tied to settlements near water holes, probably meant that malaria was nowhere hyperendemic. But by about 250 BC the Nok culture made its appearance in what is now northern Nigeria but extending both west and east at the same latitude. This belt was then probably wetter and more thickly wooded than it is today even though it was still to the north of the high forest. It was suitable for cereal growing, and the transition of people from the Stone Age into the Iron Age is thought to have taken place in that region. With farming implements made from iron, the people could cut trees and till the soil more effectively, and expand into the forest areas. These are the people who are thought to have been the original speakers of a pre-Bantu language and who subsequently migrated through (or around the eastern limits of) the rain forest to reach the Katanga and the Kasai regions of the Congo basin. Some linguists, notably Guthrie (1962), argue that it is from this pre-Bantu nucleus that the people speaking Bantu languages have originated.

With this drastic, even revolutionary, change in food production methods a veritable population explosion occurred and people settled in villages situated near convenient water sources. In such villages, people became the preferred animal for the mosquito's blood meal and thus adequate reservoirs for *P. falciparum* parasites. It would have been in this type of setting that gene mutations which conferred some degree of protection against the parasite would increase in frequency, eventually constituting a balanced genetic polymorphism. Such genetic adaptations would assist the infant in surviving the first year or two of life, by which time the child's immune system would be well on the way to producing antibodies against the malaria parasites. Although still liable to attacks of malaria, the survivors of these first crucial years would, with the help of various cultural strategies, be enabled to survive in the highly malarious environment.

The spread of iron southwards through the continent could have proceeded without the concomitant movement of large numbers of people. Similarly, Bantu languages could also have diffused without migrations of large numbers of people. But the speed with which the languages diffused (their separation is thought by linguists to date back only about 2,500 years) and the striking overall genetic similarity between all the negroid peoples of sub-Saharan Africa supports the view that significantly large numbers of people did migrate.

MALARIA-PROTECTIVE POLYMORPHISMS

The sickle cell trait

The gene responsible for sickle cell haemoglobin, $Hb\beta^S$, has a frequency of at least .08 in most of tropical Africa but south of the Zambezi River and to the south of the Cunene-Kavango Rivers the frequency of the gene drops to virtually zero. South of the Limpopo it does not occur at all.

If the populations of southern Africa are derived from immigrant Bantu-speakers who left their nuclear area of origin about 2000 years ago, it must be true that they could not have possessed $Hb\beta^S$ at that time. The gene could not have disappeared from these people in the available time span, 2000 years, even in the absence of malaria; and it is known that *P. falciparum* transmission occurs throughout the year along the coastal area as far south as Natal and in the lowveld of the Transvaal and in the bushveld areas of Swaziland (Bruce-Chwatt, 1984).

Glucose-6-phosphate dehydrogenase variants

The malaria protective allele Gd^{A-} does, however, occur in significant frequencies in all the Bantu-speaking populations of southern Africa, with frequencies ranging from a low of 0.04 in the Cape Nguni to 0.10 in the Venda of the northern Transvaal. Of course the frequency of this gene among the immigrant ancestors of these populations cannot be known - it is possible that the frequency was originally higher and, due to a relaxation in the intensity of the selective force of *P. falciparum* malaria, then declined to these values in the time span of under 2000 years. It may be significant that Gd^{A}, the G6PD allele which is associated with only 15% reduction in enzyme activity, seems to exhibit a frequency of up to about 0.15 in these same populations, similar in fact to the frequencies encountered in the Bantu-speaking populations living in hyper-endemic malaria areas further north. Perhaps the deleterious effects of Gd^{A} are much less than those of Gd^{A-} and that is why the former gene has not decreased in frequency while the latter has done so in the environment where malaria is not such a powerful selective force.

There has been no systematic search for novel G6PD variants in southern Africa apart from the one survey of Reys et al (1970) carried out in southern Mozambique. That study, in addition to establishing the frequencies of Gd^{A} and Gd^{A-} in the area, reported the presence of four probable new variants. Only variants with altered electrophoretic mobility would have been detected, so the 1.2% frequency of new variants is likely to be a gross underestimate of the true frequency. The discovery of numerous distinct deficiency variants in apparently homogeneous populations in Melanesia (Yoshida et al, 1973) should encourage a systematic search for variants in populations in tropical Africa. The use of the G6PD DNA probe (Persico et al, 1981) to characterise the DNA arrangements around the gene will, it can be predicted, prove very enlightening in such studies. For such DNA recombinant techniques will show whether all Gd^{A-} genes share a common origin, whether Gd^{A} is an older mutation than Gd^{A-}, and many more such questions.

It is now generally accepted that G6PD deficiency confers a selective advantage against *P. falciparum* malaria, even though there is still some debate over whether the female heterozygote is more highly favoured than either the hemizygote or the female homozygote. Roth et al (1983) found that parasite growth was one-third of normal in both hemi- and heterozygotes for Gd^{B-} so female heterozygotes may be at a particular advantage because they presumably have fewer risks due to G6PD deficiency than males with the variant gene. Usanga and Luzzatto (1985) showed that *P. falciparum* is able to adapt to G6PD-deficient host red cells by producing its own G6PD. This explains why Gd^{A-} males are not protected against malaria while Gd^{A-}/Gd^{B} females are. In the heterozygous female the merozoites leaving the normal red cell have a 50% chance of infecting a G6PD-deficient red cell which then has only a 50% chance of completing the next schizogonic cycle.

It is not clear, however, if the bearer of the Gd^{A-} variant allele is at any serious disadvantage - in tropical Africa or, for that matter, in more temperate regions like the United States or much of southern Africa. Favism or comparable haemolytic episodes due to naturally occurring foodstuffs have not been reported in Africans with Gd^{A} or Gd^{A-}. Drug-induced haemolytic anaemia is known but is relatively mild and self-limiting because, as younger red cells are released from the bone marrow, the level of G6PD in the circulating red cells is adequate for coping with the stressing agent (Zail et al, 1962). Bienzle (1981) confirmed that serious haemolytic episodes are very infrequently observed in medical practice in Africa; he suggested that additional red cell and serum factors, inherited independently of G6PD, might be involved in the pathogenesis of haemolytic reactions due to G6PD deficiency.

Hyperbilirubinaemia is thought to be more common in G6PD-deficient subjects during the course of certain infectious diseases than it is in non-deficient subjects: typhoid fever, lobar pneumonia and viral hepatitis have all been cited as precipitating agents. Some of the hyperbilirubinaemia may be hepatocellular in origin (Bienzle, 1981).

G6PD deficiency is thought to be an important cause of neonatal hyperbilirubinaemia in Nigeria, but Kinoti et al (1979) reported that in Nairobi, Kenya, ABO incompatibility between mother and child was the main cause of severe neonatal jaundice there. Interestingly, there is little evidence of haemolysis, and Oluboyede et al (1979) showed that the G6PD levels in the livers of Gd(-) individuals were significantly lower than the levels in matched non-deficient controls. Glucuronide production in the liver may be reduced in G6PD-deficient newborns, and Meloni et al (1973) working in Sardinia demonstrated that phenobarbital given to Gd(-) newborns could reduce the severity of hyperbilirubinaemia, thereby reducing the need for exchange transfusion. These babies would, of course, have suffered from the Mediterranean type of G6PD deficiency but the findings of Oluboyede et al (1979) suggest that a similar approach to the problem of hyperbilirubinaemia in African neonates might be successful.

Levin et al (1964) found that the African form of G6PD deficiency did render newborn babies in Johannesburg (South Africa) liable to hyperbilirubinaemia and proposed that perinatal hypoxia aggravated the condition. Of 136 G6PD-deficient newborn babies of mixed or 'Coloured' origin born in Cape Town between 1970 and 1978, 59 (43%) were shown by Roux et al (1982) to have jaundice severe enough to necessitate exchange transfusion; 18% of those suffering from idiopathic hyperbilirubinaemia needed the treatment. Kernicterus was a complication in 2.2% of G6PD deficient babies but only 2 (0.13%) of the idiopathic hyperbilirubinaemic infants developed kernicterus. The type of G6PD deficiency present in these babies was not established but it may reasonably be assumed that the overwhelming number had the Gd A(-) type. There is little harm in using G6PD-deficient blood for transfusion purposes provided that the recipient is not on large doses of drugs which cause G6PD-deficient cells to lyse. It is not considered necessary to screen blood donors for the trait.

Membrane-associated traits

Ovalocytosis is a recessively inherited condition, widely distributed in Melanesia, which has been shown to render the red cell resistant to infection by *P. falciparum*, *P. vivax* and *P. malariae* (Kidson et al, 1981). It is presumably due to an alteration in the structure of the red cell membrane, and the demonstration by Knowles et al (1984) that American blacks have a polymorphism in αII type 2 spectrin suggests that African peoples may also exhibit variation in the cytoskeleton of the red cell which may also confer an advantage against malaria.

The observation by Miller et al (1976) that Duffy negative human cells are resistant to infection by *P. vivax* stimulated a great deal of interest in "null" blood group phenotypes which may render the red cells resistant to other species of plasmodium. Pasvol and Wilson (1982) summarised a great deal of work showing that En(a-) cells are totally deficient in glycophorin A (the glycophorins are membrane sialoglycoproteins) and show a marked resistance to infection by *P. falciparum*. S-s-U- cells lack glycophorin B and show relative resistance to the parasite; if S-s-U- cells are treated with trypsin, which is known to remove glycophorin A, then total resistance is conferred on the cells. A patient with elliptocytosis, and whose red cells lacked glycophorin C, was studied by Pasvol et al (1984) and invasion of the cells by *P. falciparum* was reduced to the extent of about 57% of those of normal human red cells. Trypsin treatment of the red cells reduced invasion to 34% of normal. From these experiments it was concluded that whereas glycophorins A and C are necessary for completely successful invasion of the red cells by *P. falciparum*, glycophorin B is the only sialoglycoprotein that is essential for invasion.

It is worth considering whether malaria has been the selective agent responsible for the production of the red cell antigen polymorphisms due to glycophorin variation. There is no evidence that En(a-) individuals occur other than as 'sporadic' cases but the S^u variant allele in the MNSs system occurs at appreciable frequencies in certain African populations. Race and Sanger (1975) summarised data on the frequencies of the MS^u and NS^u alleles in two large series of US Blacks and the combined frequency is of the order of 0.10. Data on African populations, summarised in Table 1, show that the combined frequencies of MS^u and NS^u range from high values of 0.59 in Efe pygmies of the Congo region to very low frequencies in the Bantu-speaking and San populations of southern Africa.

The standard error for many of these frequencies is very high - the very presence of the allele in a population is generally only suspected if a homozygote is found - and when samples have consisted of only 100 to 200 individuals a homozygote might not have been encountered. Nevertheless, a trend is clearly apparent: the S^u allele has frequencies of around 0.20 in equatorial Africa, dropping to about 0.10 in Central Africa and reaching somewhat lower frequencies in southern Africa. It may be absent in the San but a homozygote has been found in one member of a small !Kung band at Saman!aiki. The large groups called "Other Bantu-speakers" and "Bantu-speakers" are made up of a number of smaller populations.

If the S^u allele in the MNSs system has been selected for in tropical Africa because of the advantage which it confers against *Plasmodium* malaria infection the question must be asked whether there is any disadvantage attached to carrying the gene? Heterozygotes cannot be identified except through testing relatives, but homozygous S^u/S^u individuals are known to be at risk of fatal blood transfusion reactions and women of this genotype have given birth to babies suffering from haemolytic disease of the newborn and on occasion, to a stillborn child (Race & Sanger, 1975). Fatal blood transfusions are not likely to have exerted any effect on the evolution of man, but the death of newborn babies due to materno-foetal blood group incompatibility may well have contributed to the development and maintenance of the genetic polymorphism involving the normal S and s genes and its variant S^u allele.

Table 1: The S-s-U-phenotype and allele frequencies in sub-Saharan Africa

Population	No.	S-s-U-		Allele frequency	
		No.	%	MS^u	NS^u
Nigeria[1]	121	2	1.65	0.0528	0.1074
Congo					
Babinga Pygmies[2]	146	9	6.16	0.1910	0.0852
Efe Pygmies[3]	126	44	34.92	0.5910	
Non-Pygmies[3]	660	10	1.52	0.1240	
Uganda: Karamojo[4]	108	6	5.56	0.0926	0.0781
Lacustrian Bantu[5]	138	6	4.35	0.1200	0.0883
Tanzania					
Sandawe[6]	215	2	0.93	0.0148	0.0953
Nyatura[6]	215	2	0.93	0.0211	0.0436
Malawi: Mixed[7]	204	4	1.96	0.1400	
Mozambique: Mixed[8]	569	2	0.33	0.0166	0.0407
Zimbabwe: Mixed[7]	1484	11	0.74	0.0861	
Namibia					
Ndonga (Ambo)[9]	113	1	0.88	0.0580	0.0000
Other Bantu-speakers[10]	1059	0	0.00	0.0000	0.0000
San					
Saman!aiki[9]	34	1	2.94	0.1680	0.0000
Others[9]	523	0	0.00	0.0000	0.0000
Basters[11]	120	0	0.00	0.0000	0.0000
Botswana					
Tswana[10]	192	1	0.52	0.0722	
Kgalagadi[10]	124	0	0.00	0.0000	0.0000
San[12]	230	0	0.00	0.0000	0.0000
South Africa					
Tswana[10]	109	1	0.92	0.0958	
Bantu-speakers[7,10,13]	1970	0	0.00	0.0000	

(1) Tills et al, 1979; (2) Cavalli-Sforza et al, 1968; (3) Fraser et al, 1966; (4) Allbrook et al, 1965; (5) Roberts & Ssebabi, 1976; (6) Godber et al, 1976; (7) Lowe & Moores, 1972; (8) Matznetter & Spielmann, 1969; (9) Nurse et al, 1981; (10) Jenkins (unpublished); (11) Nurse et al, 1982; (12) Nurse & Jenkins, 1977; (13) Shapiro, 1956.

The α-thalassaemia states

Before the advent of recombinant DNA technology it was not easy to assign an individual to a particular genotype at the α-globin loci. The demonstration of Hb Bart's in cord blood was probably the most sensitive technique, but this is not a very accurate method for estimating the frequency of α-thalassaemias and could hardly be applied to population studies. Globin chain synthesis studies are more accurate but need to be carried out in sophisticated laboratories, which are not readily available in Third World countries where the α-thalassaemia syndromes occur in high frequencies and where malaria is still a scourge.

The first evidence on the distribution of the α-thalassaemias in Africa came from studies using Hb Bart's levels in cord blood, which suggested that in Nigeria 5.1 to 10.7% had α-thalassaemia (Essan, 1972), in Zaire the figure was 17.9% (van Baelan et al, 1969) and in Johannesburg (South Africa) it was 1.6% (Piliszek, 1979). Haemoglobin H disease and Hb Bart's hydrops foetalis have not been described from Africa, so it has been assumed that the α^0- thalassaemia determinant (- -) does not occur; it was thought that the (-α) haplotype must occur, because of the discovery of Hb Bart's in cord blood surveys, and molecular studies have confirmed that it does. Ramsay and Jenkins (1984) have now shown that the (-α) haplotype occurs in the Venda at a frequency of 0.21 (±0.05), which is somewhat higher than the frequency found in American Blacks and West Africans. In the San its frequency is 0.06 (±0.03). In both the Venda and San populations the (-α) haplotype is of the rightward 3.7 kb deletion type. Amongst the 33 Venda, one individual was heterozygous for the triple α-globin chromosome.

The San of this study were !Kung hunter-gatherers living in the north-eastern part of Namibia. Until recently they were very isolated and detailed serogenetic studies indicate that they received little or no inflow of genes from the surrounding Bantu-speaking peoples. The Venda are Bantu-speakers living in the extreme north of the Republic of South Africa and on linguistic grounds are considered to be the most recent of the "Bantu" immigrants; they have the $Hb\beta^S$ allele at the low frequency of .002, the G6PD Gd^{A-} allele at a frequency of about .08 (higher than the frequency in the other South African chiefdoms) and Gd^A at a frequency of about .20, very similar to the frequency found throughout most of sub-Saharan Africa. The San do not have $Hb\beta^S$ or Gd^{A-} but Gd^A occurs at the low frequency of .03 to .05.

We have hypothesized that β^S is the most recent of the three well-known malaria protective alleles, Gd^A, Gd^{A-} and β^S, to have arisen in Africa and that Gd^A is the oldest (Jenkins and Dunn, 1981). The finding of the α^+-thalassaemia determinant, (-α), at such a high frequency in the Venda suggests that it is 'older' than both β^S and Gd^{A-} and that it attained these significant frequencies in the progenitors of the Bantu-speaking people before they left their nuclear area of origin in central Africa to migrate southwards. It is not certain if the α^+-thalassaemia determinant is 'older' or 'younger' than Gd^A but as its distribution becomes better known, its origin and evolution will undoubtedly be clarified.

The α-thalassaemia determinants are of little health significance in Africa. The severe pathological conditions, HbH disease and Hb Bart's hydrops foetalis, would not be expected to be encountered and this is our experience in southern Africa. Individuals possessing one (-α) determinant will not be detected in routine haematological practice but the homozygote, -α/-α, will present with a mild microcytic anaemia which is probably of no significance.

DNA variation around the β-globin gene

There is no evidence that the variation in DNA around the β-globin genes plays any part in malaria protection except in so far as the mutation might give rise to a thalassaemia syndrome. The situation of the β^S allele on the 13 kb Hpa I fragment is the general pattern in Togo, whereas in Ivory Coast only 28% of the β^S genes were on the 13 kb fragment - the other 72% were on a 7.6 kb fragment. In East African, Saudi Arabian and Indian populations the sickle cell gene is invariably on the 7.6 kb fragment.

We have investigated (Ramsay & Jenkins, 1985) the Hpa I polymorphism in the San and the Venda, among whom $Hb\beta^S$ does not, for practical purposes, occur, and have found that the 13 kb Hpa I fragment occurs in both at frequencies of .042 and .062, respectively (not a significant difference with these relatively small numbers). These frequencies are considerably lower than those found in West and East Africa, but suggest that the mutation giving rise to the 13 kb fragment is old and existed prior to the emergence of certain of the β^S mutations, in the population which was ancestral to the San and the Negro. The 7.0 kb fragment was not found in the San but it assumed a frequency of .167 in the Venda. This is higher than that encountered in any other population to date although we suspect that for technical reasons it may have gone undetected in some West African populations. At any rate, it seems to be an exclusively Negro trait, and its absence in the San suggests that it has arisen since the divergence of these two groups or else has been lost from the San; more definite interpretations must await the accumulation of more data.

CONCLUSION

This contribution has reviewed the various genetic red cell traits, both membrane-associated and intra-cellular, which are or may be responsible for conferring protection against malaria in southern Africa. Some, like Gd^{A}, probably attained their high frequencies before the progenitors of the present-day South African Bantu-speaking Negro peoples migrated into areas where malaria was not the scourge that it had been in the areas of Central Africa from which they came. It is difficult to understand how other traits, like the gene for sickle haemoglobin ($Hb\beta^{S}$), could have disappeared from these same people and it is suggested that they had migrated southwards *before* the mutation(s) had occurred. Still other traits, like Gd^{A-}, S^{u} in the MNSs system and Fy in the Duffy system, have significant frequencies, although much lower than those in Central African populations and this decrease in frequency could be due to reduction in their new environments of selection pressure for these alleles, or else to migration after the genes has arisen but before they had attained their maximal frequencies.

The disadvantageous effects of genes like Gd^{A-} and S^{u} are not easy to establish and further research is needed if they are to be incriminated as causes of ill health.

ACKNOWLEDGMENTS

We are grateful to Coleen Morgan, David Dunn and Shirley Alper for their help in the preparation of this paper.

REFERENCES

Allbrook, D., Barnicot, N.A., Dance, N., Lawler, S.D., Marshall, R. & Mungai, J. (1965). Blood groups, haemoglobin and serum factors of the Karamojo. Human Biology, **37**, 217-237.

Bienzle, U. (1981). Glucose-6-phosphate dehydrogenase deficiency, Part I: Tropical Africa. In: L. Luzzato (ed.), Clinics in Haematology, **10**, 785-799.

Bruce-Chwatt, L.J. (1965). Paleogenesis and paleo-epidemiology of primate malaria. Bulletin of the World Health Organization, **32**, 365-387.

Bruce-Chwatt, L.J. (1984). Lessons learned from applied field research activities in Africa during the malaria eradication era. Bulletin of the World Health Organization, **62** (Suppl.), 19-29.

Cavalli-Sforza, L.L., Zonta, L.A., Nuzzo, F., Bernini, L., De Jong, W.W.W., Meera Khan, P., Ray, A.K., Went, L.N., Siniscalco, M., Nijenhuis, L.E., van Loghem, E. & Modiano, G. (1969). Studies on African pygmies. I. A pilot investigation of Babinga pygmies in the Central African Republic (with an analysis of genetic distances). American Journal of Human Genetics, **21**, 252-274.

Essan, F.G.S. (1972). Haemoglobin Bart's in newborn Nigerians. British Journal of Haematology, **22**, 73-80.

Fraser, G.R., Giblett, E.R. & Motulsky, A.G. (1966). Population genetic studies in the Congo. iii. Blood groups (ABO, MNSs, Rh, Js[a]). American Journal of Human Genetics, **18**, 546-552.

Godber, M., Kopec, A.C., Mourant, A.E., Teesdale, P., Tills, D., Weiner, J.S., El-Niel, H., Wood, C.H. & Barley, S. (1976). The blood groups, serum groups, red cell isoenzymes and haemoglobins of the Sandawe and Nyaturu in Tanzania. Annals of Human Biology, **3**, 463-473.

Guthrie, M. (1962). Some developments in the prehistory of the Bantu languages. Journal of African History, **3**: 273-282.

Jenkins, T. & Dunn, D. (1981). Haematological genetics in the tropics. Clinics in Haematology, **10**, 1029-1049.

Kidson, C., Lamont, G., Saul, A. & Nurse, G.T. (1981). Ovalocytic erythrocytes from Melanesians are resistant to invasion by malaria parasites. Proceedings of the National Academy of Sciences, U.S.A., **78**: 5829-5832.

Kinoti, S.N., Kimenmiah, S.G. & Hutcheon, R.A. (1979). The incidence of ABO haemolytic disease of the newborn at Kenyatta National Hospital, Nairobi, 1977. South African Medical Journal, **56**, 127-132.

Knowles, W.J., Bologna, M.L., Chasis, J.A., Marchesi, S.L. & Marchesi, V.T. (1984). Common structural polymorphisms in human erythrocyte spectrin. Journal of Clinical Investigations, **73**, 973-979.

Levin, S.E., Charlton, R.W. & Freiman, I. (1964). Glucose-6-phosphate dehydrogenase deficiency and neonatal jaundice in South Arican Bantu infants. Journal of Pediatrics, **65**, 757-764.

Lowe, R.F. & Moores, P.P. (1972). S-s-U- red cell factor in Africans of Rhodesia, Malawi, Mozambique and Natal. Human Heredity, **22**, 344-350.

Matznetter, T. & Spielmann, W. (1969). Blutgruppen Mocambiquanischer Bantustammer. Zeitschrift fur Morphologie und Anthropologie, **61**, 57-71.

Meloni, T., Cagnazzo, G., Dore, A. & Cutillo, S. (1973). Phenobarbital for prevention of hyperbilirubinaemia in glucose-6-phosphate dehydrogenase deficient newborn infants. Journal of Pediatrics, **82**, 1048-1051.

Miller, C.H., Mason, S.J., Clyde, D.F. & McGinniss, M.H. (1976). The resistance factor to *Plasmodium vivax* in Blacks. The Duffy blood group genotype, *Fy Fy*. North of England Journal of Medicine, **295**, 302-304.

Nurse, G.T. & Jenkins, T. (1977). Health and the Hunter-Gatherer. Basel: Karger.

Nurse, G.T., Jenkins, T., Africa, B.J. & Stellmacher, F.F. (1982). Sero-genetic studies on the Basters of Rehoboth, South West Africa/Namibia. Annals of Human Biology, **9**, 157-166.

Nurse, G.T., Rootman, A.J. & Jenkins, T. (1981). Sero-genetic studies on the Ambo of Namibia (in preparation).

Oluboyede, O.A., Esan, G.J.F., Francis, T.I. & Luzzatto, L. (1979). Genetically determined deficiency of glucose-6-phosphate dehydrogenase (type A^-) is expressed in the liver. Journal of Laboratory Clinical Medicine, **93**, 783-789.

Pasvol, G. & Wilson, R.J.M. (1982). The interaction of malaria parasites with red blood cells. British Medical Bulletin, **38**, 133-140.

Pasvol, G., Anstee, D. & Janner, M.J.A. (1984). Glycophorin C and the invasion of red cells by *Plasmodium falciparum*. Lancet, **2**, 907-908.

Persico, M.G., Toniolo, D., Nobile, C., d'Urso, M. & Luzzatto, L. (1981). cDNA sequences of human glucose-6-phosphate dehydrogenase cloned in pBR 322. Nature, **294**, 778-780.

Piliszek, T.S. (1979). Hb Bart's and its significance in the South African Negro. Acta Haematologica, **61**, 33-38.

Race, R.R. & Sanger, R. (1975). Blood Groups in Man, 6th edn. London: Blackwell Scientific Publications.

Ramsay, M. & Jenkins, T. (1984). α-Thalassaemia in Africa: the oldest malaria protective trait? Lancet, **2**, 410.

Ramsay, M. & Jenkins, T. (1985). The Hpa I β-globin restriction fragment length polymorphism in Southern Africa. South African Journal of Science, **81**, 98-100.

Reys, L., Manso, C. & Stamatoyannopoulos, G. (1970). Genetic studies on Southeastern Bantu of Mozambique. I. Variants of glucose-6-phosphate dehydrogenase. American Journal of Human Genetics, **22**, 203-215.

Roberts, D.F., Ssebabi, E.C.T. (1976). East African frequencies in some lesser known blood group systems. Man, 10, 524-529.

Roth, E.F., Raventos-Suarez, C., Rinaldi, A. & Nagel, R. (1983). Glucose-6-phosphate dehydrogenase deficiency inhibits *in vitro* growth of *Plasmodium falciparum*. Proceedings of the National Academy of Sciences, U.S.A., **80**, 298-299.

Roux, P., Karabus, C.D. & Hartley, P.S. (1982). The effect of glucose-6-phosphate dehydrogenase deficiency on the severity of neonatal jaundice in Cape Town. South African Medical Journal, **61**, 781-782.

Shapiro, M. (1956). Inheritance of the Henshaw (He) blood factor. Journal of Forensic Medicine, **3**, 152-160.

Ssebabi, E.C.T. & Roberts, D.F. (1976). East African frequencies in some lesser known blood group systems. Man (N.S.), **10**, 524-529.

Tills, D., Kopec, A.C., Fox, R.F. & Mourant, A.E. (1979). The inherited blood factors of some Northern Nigerians. Human Heredity, **29**, 172-176.

Usanga, E.A. & Luzzatto, L. (1985). Adaptation of *Plasmodium falciparum* to glucose-6-phosphate dehydrogenase-deficient host red cells by production of parasite-encoded enzyme. Nature, **313**, 793-795.

Van Baelan, H., Van de Pitte, J., Cornu, G. & Beckles, R. (1969). Routine detection of sickle cell anaemia and haemoglobin Bart's in Congolese neonates. Tropical Geogr. Medicine, **21**, 412-415.

Yoshida, A., Giblett, E.R. & Malcolm, L.A. (1973). Heterogeneous distribution of glucose-6-phosphate dehydrogenase variants with enzyme deficiency in the Markham Valley area of New Guinea. Annals of Human Genetics, **37**, 145-150.

Zail, S.S., Charlton, R.W. & Bothwell, T.H. (1962). The haemolytic effect of certain drugs in Bantu subjects with a deficiency of glucose-6-phosphate dehydrogenase. South African Journal of Medical Science, **27**, 95-99.

THE GENETIC ORIGIN OF THE VARIABILITY OF THE PHENOTYPIC EXPRESSION OF THE Hb S GENE

D. LABIE[1], J. PAGNIER[1], H. WAJCMAN[1], M. E. FABRY[1] and R.L. NAGEL[2]

[1] *Institut de Pathologie Moleculaire, INSERM, Paris, France*
[2] *Division of Hematology, Albert Einstein College of Medicine, New York, U.S.A.*

INTRODUCTION

All individuals with sickle cell anemia have two β genes that encode a valine residue instead of a glutamic residue in position 6 (Bookchin & Nagel, 1981). Nevertheless the clinical or phenotypic expression of this homozygous genetic abnormality is extremely variable from individual to individual (Labie, 1983; Nagel & Fabry, 1986). Although environmental factors are obviously involved they do not account for most of the variability, and other genetic factors have to be invoked. In the last few years significant progress has been made in this area and the findings from our laboratories and those of others indicate three genetic sources for the variability in the expression of Hb S homozygosity:

SOURCES OF VARIABILITY

Derived from multiple gene origin: sickle cell anemia is a multihaplotype disease

If the Hb S gene had originated only once in history and then expanded in frequency, all sickle cell anemia patients not only would be homozygous for the mutation but would tend to have identical genes and other sequences surrounding the mutated β globin gene. Even if genes or sequences to the left or to the right of the β globin gene could modify the expression of the Hb S gene, a single origin would guarantee that a significant and identical genomic block around this gene would be inherited as an entity by all individuals carrying the Hb S gene. Hence, even in the presence of modifying or epistatic genes located nearby, the variability of phenotypic expression would be minimal unless hot spots for recombination are present. In the opposite situation, if the mutant gene originated more than once, the heterogeneity of origin would make it improbable that the 5' and 3' regions around the Hb S gene would be identical in each instance. This low probability derives from the high frequency of polymorphic sites involving genes and flanking regions that characterise the β-like globin cluster, a situation probably very common elsewhere in the genome.

The latter situation is indeed the case for the Hb S mutation. The gene seems to have originated in three distinct geographical areas of Africa, and in each instance the genome surrounding the mutated β globin gene is distinctly different as judged by endonuclease-detectable polymorphic sites (Pagnier et al, 1984) (Figs. 1 and 2). This circumstance is capable of introducing phenotypic variation in sickle cell anemia. Linked to the β gene are the two structural γ genes (and presumably their controlling sequences) involved in the synthesis of Hb F, a protein that significantly influences the tendency of Hb S to polymerise (Bookchin et al, 1975; Nagel et al, 1979). Recently we have compared the hematological and genetic characteristics of sickle cell patients in Senegal and Benin. The differences are depicted in Table 1, and these data demonstrate abundantly the capacity of multiple gene origin to influence the phenotypic expression of a genetic disease.

Derived from unlinked gene epistasis: sickle cell anemia is a multigene disease

Another source of variability is the influence of unlinked epistatic genes. A very good candidate from early on was the α locus on chromosome 16. Deletions of the duplicated α gene loci generate two forms of α thalassaemia in black populations: -α/αα (α 3) and -α/-α (α 2). Several authors (Embury et al, 1982; Higgs et al, 1982) provided evidence that α thalassaemia ameliorates the hematological picture of sickle cell anemia. Nevertheless, the effect is not a simple one, as other features of the disease become more frequent (Table 2). On balance, the very high frequency of α thalassaemia among Blacks and carriers of the Hb S gene makes this source of variability a very important and quantitatively significant one.

This effect emphasizes the multigenic character of sickle cell anemia. All patients are similarly homozygous for the mutated β globin gene but they must have a myriad of other unlinked genes that affect differentially several of the clinical features of the disease. A future area of enquiry must be genes involved in the regulation of the tone of the microcirculation and the response to initial vaso-occlusion. As the properties of sickle cells are necessary but not sufficient to explain the incidence of painful crises in this disease (Nagel & Fabry, 1986), modulation at the level of the microcirculation must make an

Table 1: Hematological and genetic data of SS patients from two haplotypically homogeneous areas of Africa †

	% Hb F	Gγ	F4	ISC
Senegal	↑	↑	↓	↓
Benin	↓	↓	↑	↑

† Nagel et al, 1985.

Gγ: the Gγ gene expression as a percentage of fetal γ chains.

F4: densest red cell fraction by isopyknic gradient.

ISC: irreversible sickle cell count.

Hb F: percentage of fetal hemoglobin.

Table 2: Effects of the concomitant presence of α thalassemia (-α/αα) on sickle cell anemia (†)

Hb and Hct	↑
MCHC	↓
Survival	↑
Dense red cells	↓
Splenic sequestration	↑
Aseptic vascular necrosis	↑

† References:

Embury et al, 1982, 1984; Higgs et al, 1982; Mears et al, 1982, 1983; Fabry et al, 1984.

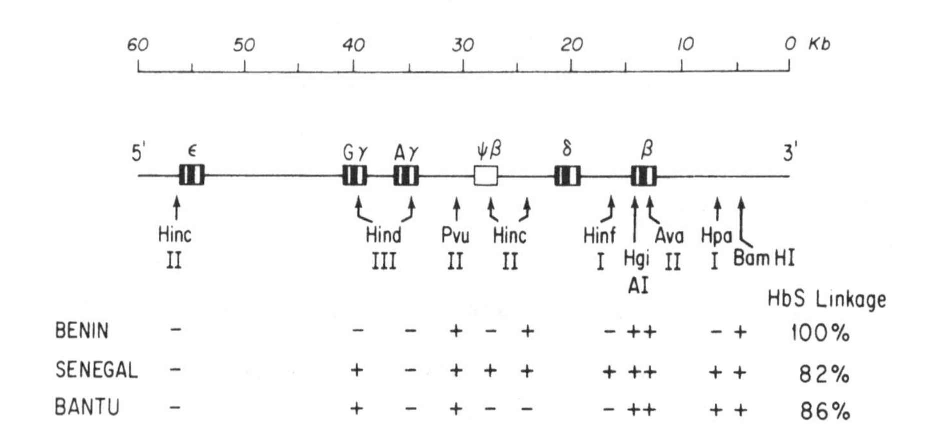

Figure 1: The three main Hb S-linked haplotypes in Africa.

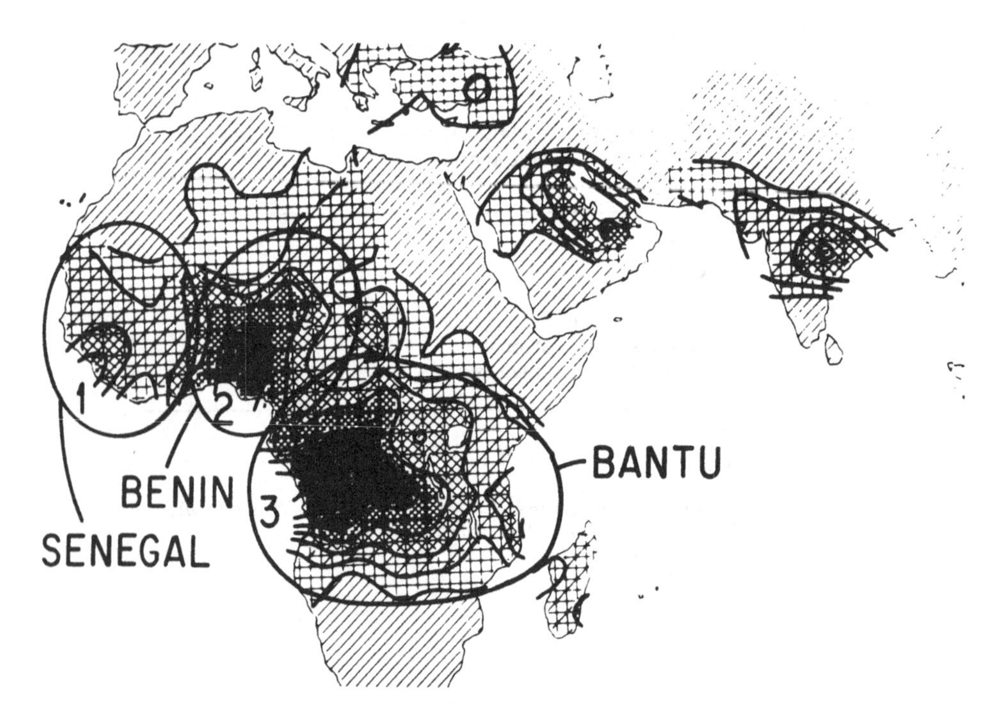

Figure 2: Approximate borders of haplotype homogeneous areas of Africa

important contribution to this devastating frequent complication.

Derived from gene flow: sickle cell anemia is genetically and hematologically different in various geographic locations

The hemoglobin S gene originated approximately between 3000 and 2000 years ago in (at least) three different and distinct geographical areas of Africa (Nagel & Labie, 1985). Considerable gene flow coupled with the selective pressure of malaria distributed, and increased the frequency of, these three mutations throughout vast areas of Africa (Atlantic West Africa for the Senegal gene, Central West Africa for the Benin gene, and the Bantu speaking Equatorial, eastern and southern Africa for the gene identified by us in the Central African Republic). This gene flow was probably helped by the ancient African custom of making slaves of defeated neighbours but incorporating their descendants as fully fledged members of the community, sometimes even attaining exalted social positions. In other words, slavery lasted only one generation. Slavery was also practised in Europe at the time, but in a less generous fashion.

Another less important instance of gene flow was probably involved in the dispersion of the Bantu Hb S gene (Fage, 1979). About 2000 years ago, the populations living along the Benue River in eastern Nigeria expanded towards the east and the south of Africa, engulfing and largely destroying the Pygmy and Bushman communities pre-existing in those areas. Remnants of these peaceful non-Negro ethnic groups can be found today in the Central African Republic and Cameroon (Pygmies) and in the Kalahari desert of southwest Africa (Bushman).

This distribution of the several Hb S genes in Africa has to be considered in conjunction with the variation in frequency of a non-linked epistatic gene deletion, α thalassemia. For example, in Senegal α thalassemia frequency is one-third (0.10) that in Benin (0.30) (Pagnier et al, 1984), adding to the extent and geographical diversity of the variability.

It is of interest that the Hb S gene found in North Africa is of the Benin type, strongly suggesting that this gene migrated from Central West Africa along the well-known trans-Saharan commercial routes to North Africa and probably to the rest of the Mediterranean. Interestingly enough, the deletional α thalassemia gene either did not arrive simultaneously with the Hb S gene (maybe due to differences in the time of origin) or was eliminated from the population through the potential advantage of the non-deletional α thalassemia found among the North African populations.

The Americas are a fascinating example of the diversity of gene flow. All the Hb S genes present in the Americas must have come from Africa, on account of the complete absence of this gene in the pre-colonial Amerindian population. Nevertheless, from one area of the Americas to another there must be enormous variation of the different Hb S linked-haplotype frequencies and probably in the incidence of α thalassemia. This diversity derives from the many changes, during the 400 years of slave trade, in the ports of origin of the slaves brought to the Americas depending on what European power was involved in the trading and the colony for which this human traffic was destined. A remarkable example of this phenomenon is provided by the differences in frequency of the Senegal haplotype in parts of the U.S. from those expected according to the records of the British slave trade (Curtin, 1969). As Senegalese slaves were in particularly high demand in South Carolina (the slave market that "supplied" the south of the present day United States), few individuals of this origin were distributed through the northern market in Jamesville (present day Virginia). A low frequency of the Senegalese haplotype characterises sickle cell anemia patients in Maryland (Antonarakis, 1984).

In conclusion, the variability of the homozygosity for Hb S is the consequence of a number of genetic factors. These include the multiplicity of gene origins, with their corresponding different haplotypes and "fellow-traveller" genes; the presence of at least one (and probably many other) epistatic non-linked genes; and the vagaries of gene flow that mixed and rearranged the frequencies of the original β^S linked-haplotypes and other genes involved. This situation has created at least three layers of genetic variability, explaining the striking diversity in individual and geographically-specific phenotypic expression of sickle cell anemia.

REFERENCES

Antonarakis, S. E., Boehm, C. D., Serjeant, G. R., Theisen, C. E., Dover, G. J. & Kazazian, H. H. Jr. (1984). Origin of the βS Globin Gene in Blacks. The contribution of recurrent mutation and/or gene conversion. Proceedings of the National Academy of Science, U.S.A., **81**, 853-856.

Bookchin, R. M. & Nagel, R. L. (1981). Molecular and Cellular Aspects of Red Cell Sickling. In: R. Silber (ed.), Contemporary Hematology/Oncology vol. 2, pp. 31-77. Plenum Press.

Bookchin, R. M., Nagel, R. L. & Balazs, T. (1975). Gelation of hemoglobin S: role of hybrid tetramer formation. Nature, **256**, 667-668.

Curtin, P. D. (1969). The Atlantic Slave Trade. A Census. The University of Wisconsin Press, Milwaukee.

Embury, S. H., Clark, M. R., Monroy, G. & Mohandas, N. (1984). Concurrent sickle anemia and α-thalassemia. Effect on pathological properties of sickle erythrocytes. Journal of Clinical Investigation, **73**, 116-123.

Embury, S. H., Dozy, A. M., Miller, J., Davis, J. R. Jr., Kleman, K. M., Preisler, H., Vichinsky, E., Lande, W. N., Lubin, B. H., Kan, Y. W. & Mentzer, W. C. (1982). Concurrent sickle-cell anemia and α-thalassemia: effect on severity of anemia. New England Journal of Medicine, **306**, 270-274.

Fabry, M. E., Mears, J. G., Patel, P., Schaefer-Rego, K., Carmichael, L. D., Matinez, G. & Nagel, R. L. (1984). Dense cells in sickle cell aemia: the effects of gene interaction. Blood, **64**, 1042-1046.

Fage, J. D. History of Africa. New York: Knoft.

Hawker, H., Neilson, H., Hayes, R. J. & Serjeant, G. R. (1982). Haematological factors associated with avascular necrosis of the femoral head in homozygous sickle cell disease. British Journal of Haematology, **50**, 29-34.

Higgs, D. R., Aldridge, B. E., Lamb, J., Clegg, J. B., Weatherall, D. J., Hayes, R. J., Grandison, Y., Lowrie, Y., Mason, K. P., Serjeant, B. E., & Serjeant, G. R. (1982). The interaction of α-thalassamia and homozygous sickle-cell disease. New England Journal of Medicine, **306**, 1441-1446.

Labie, D. (1983). Facteurs genetiques d'attenuation de la drepanocytose. Bull. Eur.Physiopathol.Respir., **19**, 272-273.

Mears, J. G., Lachman, H. M., Labie, D. & Nagel, R. L. (1983). α-thalassemia is related to prolonged survival in sickle cell anemia. Blood, **62**(2), 286-291.

Mears, J. G., Schoeburn, M., Schaefer, K. E., Bestak, M. & Radel, E. (1982). Frequent association of α-thalassemia with splenic sequestration crisis and splenomegaly in sickle cell (SS) subjects. Blood, **60**, 47a.

Nagel, R. L., Bookchin, R. M., Johnson, J., Labie, D., Wajcman, H., Isaac-Sodeye, A. W., Honing, G. R., Schiliro, G., Crookston, J. H. & Matsumo, K. (1979). The structural bases of the inhibitory effects of Hb F and Hb A_2 on the polymerization of Hb S. Proceedings of the National Academy of Science, U.S.A., **76**, 670-672.

Nagel, R. L. & Fabry, M. E. (1986). The many pathophysiologies of sickle cell anemia. American Journal of Hematology.

Nagel, R. L. & Labie, D. (1985). The consequence and implication of the multicentric origin of the Hb S gene, In: G. Stamatoyannopoulos & A. W. Nienhuis (eds.), Experimental Approaches for the Study of Hemoglobin Switching, pp. 93-103. Liss, New York.

Nagel, R. L., Fabry, M. E., Pagnier, J., Zohoun, I., Wajcman, H., Baudin, V. & Labie, D. (1985). Hematologically and genetically distinct forms of sickle cell anemia in Africa: the Senegal type and the Benin type. New England Journal of Medicine, **312**, 880-884.

Pagnier, J., Dunda-Belkhodja, O., Zohoun, I., Teyssier, J., Baya, H., Jaeger, G., Nagel, R. L. & Labie, D. (1984). α-thalassaemia among sickle cell anemia patients in various African populations. Human Genetics, **68**, 318-319.

Pagnier, J., Mears, J. G., Dunda-Belkhodja, O., Schaefer-Rego, K. E., Beldjord, C., Nagel, R. L. & Labie, D. (1984). Evidence of the multicentric origin of the hemoglobin S gene in Africa. Proceedings of the National Academy of Science, U.S.A., **81**, 1771-1773.

ORIGIN AND MAINTENANCE OF GENETIC VARIATION IN BLACK CARIB POPULATIONS OF ST. VINCENT AND CENTRAL AMERICA

M. H. CRAWFORD

Department of Anthropology, University of Kansas
Lawrence, Kansas, U.S.A.

INTRODUCTION

The Black Caribs of Central America (also known as the *Garifuna*) provide a tropical example of an evolutionarily successful colonising population. They increased from fewer than 2000 persons in the year 1800 to more than 80,000 in less than 180 years. Primarily through population fission, the five founding Black Carib communities established a total of 54 coastal towns and villages. This rapid geographical expansion from the Bay of Honduras to Guatemala, British Honduras and Nicaragua extended over approximately 1000 kilometers of Central American coastline (Figure 1).

The successful colonisation of the coast by the Black Caribs can be explained in part by: (1) an exceptional level of genetic variation contained in the gene pool of the founding group; (2) pre-existing genetic adaptation against malaria; (3) a social organisation which maintained high level of fertility and genetic variation; (4) a unique sequence of historical events. This recent expansion of the Black Caribs provides an excellent opportunity to explore the dynamics and population structure of tropical colonising populations.

HISTORICAL BACKGROUND

The Black Caribs originated on St. Vincent Island of the Lesser Antilles as an amalgam between Carib/Arawak Amerindians and West Africans (Figure 1). Initially, St. Vincent Island was settled by Arawaks from South America in approximately 100 A.D. Between 1200 A.D. and the time of European contact, another Amerindian group, the Caribs, expanded from Venezuela and intermixed with the original inhabitants of St. Vincent. Archaeological evidence suggested a massive invasion by the Caribs, exterminating the Arawak males but hybridising with the Arawak females (Rouse, 1976). However, Gullick (1979) argued in favour of a relatively small number of Carib-speaking Amerindians diffusing into

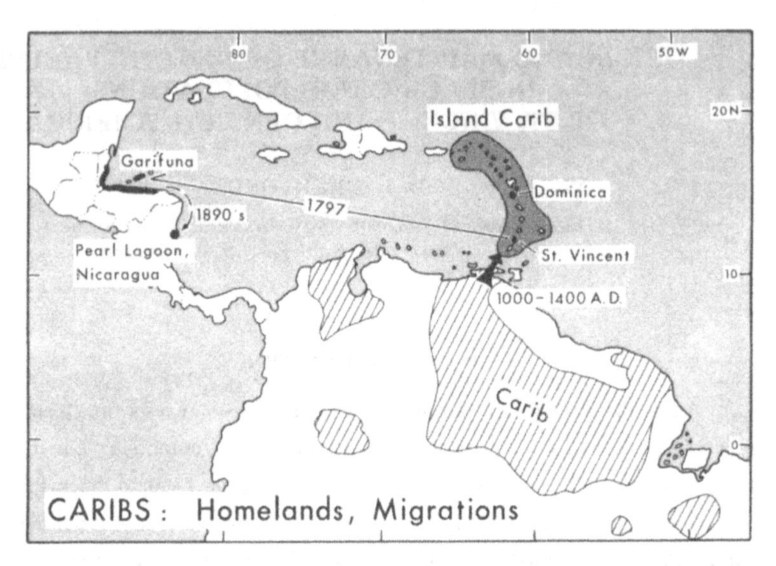

Figure 1: The migration patterns of the Carib Indians from South America to the Caribbean, followed by their hybridization with Africans on St. Vincent Island, and their relocation to Roatan and Central America. (From Davidson, 1984.)

the Caribbean and hybridising with the Arawaks. The linguistic evidence supports Gullick's position in that Island Carib (the language spoken by the Black Caribs) consists primarily of Arawakan, but with Carib, French, English and Spanish loan words.

From 1517 to 1646, the African component was added to the Black Carib gene pool. A number of explanations have been proposed as to the source of this African admixture: (1) runaway slaves from Barbados, an adjoining island and a centre of the Caribbean slave trade; (2) Amerindian raids of adjoining European settlements, returning with African captives; (3) shipwreck of an 18th century slave galleon that was heading for Barbados. This last explanation appears to be most unlikely because, if the slaver had sailed directly from Africa, then the culture of the survivors of this galleon should have been African orientated; yet, the slaves had acquired Arawak culture and language with little or no evidence of significant African influence (Taylor, 1951).

Following the Treaty of Paris in 1753, the British annexed St. Vincent Island from the French (Gulllick, 1984). However the Black Caribs sided with the French, and resisted British colonisation by burning plantations and leading an armed insurrection. After prolonged warfare with the British expeditionary military force, the surviving Black Caribs were rounded up and transported to the adjoining island, Balliceaux. During their internment on this island, the Black Carib numbers were reduced from 4200 to 2026 by an epidemic, possibly typhus (Gonzalez, 1984). In 1797, these survivors were shipped to Roatan (one of the Bay Islands, off the coast of Honduras). From there the Spanish colonial government arranged for their resettlement in the Bay of Honduras.

On St. Vincent Island, a small group of Black Caribs managed to avoid capture and they were the founders of the contemporary Black Carib population. In 1805, this founding group numbered 16 males, 9 females and 20 children (Gullick, 1976). The population of Black Caribs on St. Vincent gradually grew over the succeeding 180 years with occasional genetic bottlenecks resulting from emigrations and the high mortality associated with two eruptions of the Soufriere volcano. At present, Gullick (1984) estimates a total of 5000 Black Carib residents on the island. However, a recent government census enumerated only 2000 Black Caribs located in the north Windward section of the island (Figure 2).

The majority of Black Caribs that were transplanted to Honduras established five villages along the Bay of Honduras, near the present city of Trujillo. In 1803-4, almost immediately after their arrival in Honduras some of the Garifuna moved into an area east of Mosquitia (Davidson, 1984). In successive migrations the Black Caribs settled in British Honduras, Guatemala, Western Honduras and even Nicaragua (Figure 3). Political intrigues, population pressures, and a scarcity of cultivable land rapidly dispersed the Black Caribs along 1000 kilometres of the coast of Central America into the present 54 distinct communities (Figure 4).

METHODS

During the summers of 1975, 1976 and 1979, the University of Kansas field research team collected a total of 1044 blood specimens from Black Caribs residing in Central America and St. Vincent Island (Table 1). The blood specimens were sent to the Minneapolis War Memorial Blood Bank and the American National Red Cross for analyses of blood groups, serum and red blood cell protein types and the allele frequencies were computed by maximum likelihood.

Following examination of their genetic variation, the genetic structure of the Black Caribs of the coast of Central America and the Caribbean was

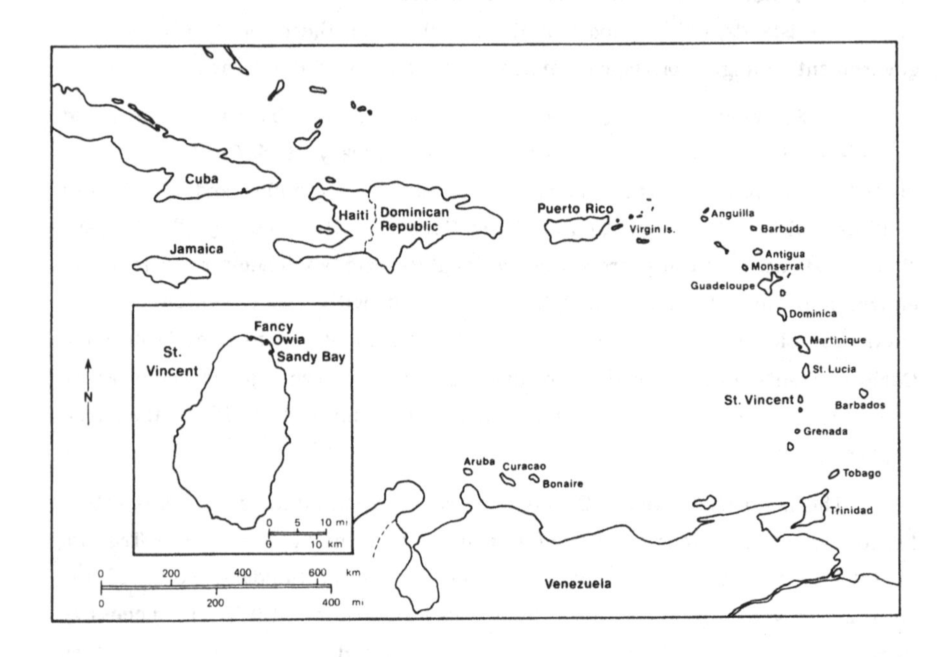

Figure 2: A map locating St. Vincent Island in the Lesser Antilles and the Black Carib communities of Fancy, Owia and Sandy Bay.

examined through a number of analytical techniques. These are: (1) R-matrix of conditional kinship; (2) admixture estimates; (3) computation of absolute heterozygosity; (4) the relationship between mean per locus heterozygosity (H_i) and r_{ii} (the genetic distance from the centroid of distribution). The R-matrix method of analysis was first developed by Harpending and Jenkins (1973) and has been widely utilised to construct "genetic maps" of subdivided populations based upon a least-squares approximation of a variance-covariance matrix. The

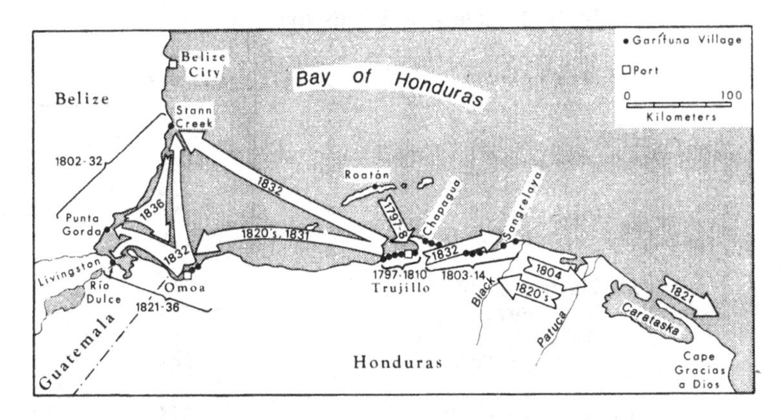

Figure 3: Migration patterns of the Black Caribs after their arrival on the coast of Honduras. (From Davidson, 1984.)

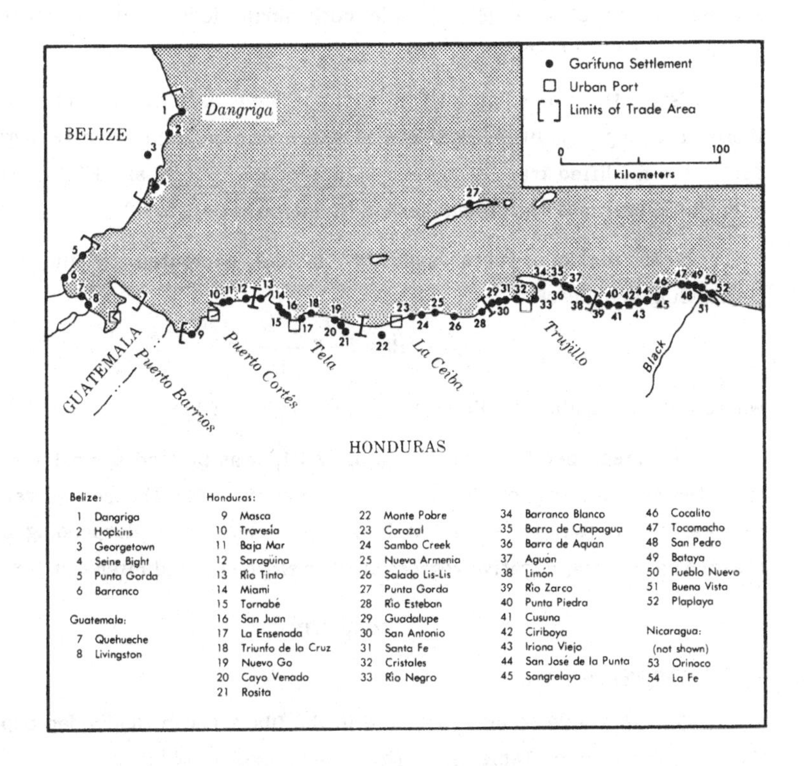

Figure 4: The location of the 54 Black Carib communities of Central America. (From Davidson, 1984.)

Table 1: Sample sizes in this study

Population	Sample size (n)
Mainland samples:	
Livingston	205
Stann Creek (Dangriga)	354
Punta Gorda	239
Total Caribs	798
St. Vincent samples:	
Sandy Bay	161
Owia	85
Total Caribs	246
Total Caribs in study	1044

admixture estimates utilised in this analysis, described in Crawford et al (1984) and Schanfield et al (1984) include both single locus and an assortment of multiple loci estimates.

Estimations of admixture for a single locus used Bernstein's formula, and those based upon multiple loci utilise the maximum likelihood (Krieger et al, 1963), the modified true least squares method of Roberts and Hiorns (1965), and a weighted multiple regression method (Crawford et al, 1976).

Mean per-locus heterozygosity $\bar{H}_i$ was computed by the method of Harpending and Chasko:

$$\bar{H}_i = 1 - \Sigma \frac{p^2 k}{K}$$

where k is the number of alleles and K the number of loci.

The mean per-locus heterozygosity (H_i) was plotted against the distance from the centroid (r_{ii}) of the relationship matrix. The theoretical relationship between these two variables has been demonstrated by Harpending and Ward (1982) as a uniform negative regression, with slope -Hp and intercept Hṗ.

RESULTS

Genetic variation

Out of the 29 blood loci tested in the Black Carib subdivided populations, 26 are polymorphic (Table 2). The three monomorphic are malate dehydrogenase (MDH), lactate dehydrogenase (LDH) and phosphoglucomutase (PGM_2).

Table 2: Polymorphic loci among the Black Caribs

Loci	Alleles with an incidence of 1% or greater
BLOOD GROUPS	
ABO	A1, A2, B, O
Rhesus	CDE, CDe, cDE
MNSs	MS, Ms, NS, Ns, S^u, Mg
Duffy	Fy^a, Fy^b, Fy
Diego	Di^a, Di^b
Kidd	Jk^a, Jk^b
Henshaw	He+, He-
Kell	K, k
P	P+, P-
Lutheran	Lu^a, Lu^b
Js	Js^a, Js^b
SERUM PROTEINS	
Albumin	$A1^{Me}$, $A1^A$
Haptoglobins	Hp^1, Hp^2, Hp^{1M}
Transferrins	Tf^C, Tf^D, Tf^{D1}
Ceruloplasmin	Cp^A, Cp^B
Group-specific component	Gc^{1S}, Gc^{1F}, Gc^2, Gc^{Ab}
Gamma globulins	$Gm^{z,a;b}$, $Gm^{z,a;b,c3,5}$, $Gm^{z,a;b,c3}$, $Gm^{z,abs}$, $Gm^{f;b}$, Gm^{zag}, Gm^{zaxg}
Km	Km^1, Km^2
Properdin factor B (Bf)	Bf^S, Bf^F, $Bf^{S0.7}$, Bf^1
RED CELL PROTEINS	
Acid phosphatase	P^A, P^B, P^C, P^R
Esterase D	EsD^1, EsD^2
Adenylate kinase	AK^1, AK^2
6-phosphogluconate dehydrogenase	PGD^A, PGD^C
Adenosine deaminase	ADA^1, ADA^2
Phosphoglucomutase IEF	PGM_i^{1+}, PGM_i^{1-}, PGM_i^2, PGM_i^{2-}
Glucose-6-phosphate dehydrogenase	Gd^A, Gd^{A-}, Gd^B

Tables 3, 4 and 5 summarise the blood group, serum and red cell protein gene frequencies utilised in the R-matrix analysis. Complete phenotypic and allelic frequencies for these Black Carib populations are available elsewhere (Crawford et al, 1981, 1984; Crawford, 1983).

The incidence of the various blood markers on St. Vincent and the Central American Black Carib populations reflect the complex history of these groups. For example, the lowest incidence of the cDe(Ro) haplotype of the rhesus system in Garifuna populations occurs in Sandy Bay, 0.247, by comparison with figures of 0.418 to 0.566 among the other populations (Table 3). This low value reflects the considerable Amerindian contribution and low African admixture in the Sandy Bay founders. Similarly, both CDe(R1) and cDE(R2) Amerindian markers have high frequencies in the Black Carib populations, with the highest occurring in Sandy Bay.

The presence of albumin Mexico in the Central American Black Carib populations and its absence on St. Vincent Island, suggest a Central American origin of the gene. The A1M variant could not have been introduced by the Carib or Arawak Indians into this gene pool, because it is absent in South American indigenous populations. The economic and social contacts between the Maya Indians of Belize and the Black Caribs and Creoles of Stann Creek, Punta Gorda and Belize City have resulted in some gene flow between these groups. The greater isolation of the Livingston and Hopkins Black Carib communities from Amerindian groups would explain their lower incidence of the albumin Mexico variant (Table 4).

While the Gc Aborigine variant is absent among the Black Carib groups of Central America, it occurs at a range of 1.2 to 2.8% among the Black Caribs of St. Vincent Island (Table 4). Considering the geographical distribution of this allele, it is likely that when the St. Vincent Carib gene pool was subdivided by the British in 1797, fortuitously those individuals with GcAb escaped deportation and founded the communities of Sandy Bay, Owia and Fancy (Crawford et al, 1984).

The levels of genetic varition estimated by mean heterozygosity per locus for the St. Vincent Island Caribs closely resembled those observed in coastal Central American communities, with H_i = 46.2% and 45% respectively. This level of genetic variation is high when compared to African groups such as the !Kung Bushman where H_i ranges from 26% to 34% in a series of populations (Harpending and Chasko, 1976).

Table 3: Blood group gene frequencies in Black Carib populations

System and Allele	Belize		St. Vincent		Guatemala
	Stann Creek	Punta Gorda	Owia	Sandy Bay	Livingston
ABO:					
A_1	0.110	0.077	0.033	0.056	0.064
A_2	0.042	0.030	0.076	0.095	0.032
B	0.129	0.092	0.109	0.086	0.160
O	0.719	0.801	0.782	0.763	0.744
Rhesus:					
CDE	0.016	0.016	0.000	0.000	0.034
CDe	0.119	0.000	0.112	0.166	0.091
cDE	0.172	0.189	0.286	0.314	0.145
cDe	0.418	0.456	0.450	0.247	0.508
Cde	0.028	0.099	0.000	0.000	0.121
cde	0.247	0.240	0.134	0.269	0.101
CdE	0.000	0.000	0.019	0.004	0.000
MNSs:					
MS	0.136	0.189	0.075	0.052	0.110
Ms	0.346	0.322	0.328	0.388	0.314
NS	0.108	0.071	0.170	0.087	0.108
Ns	0.410	0.418	0.438	0.473	0.468

There is considerable variability at the Black Carib gamma globulin (Gm) locus. In comparison to the European populations, with only 10 to 12 distinct Gm phenotypes, 28 and 42 distinct phenotypes occur in the Black Carib population of St. Vincent and Belize respectively (Schanfield et al, 1984). The high level of variation at this locus is due, in part, to the initial hybridisation between two Amerindian populations (Arawak and Carib), followed by gene flow from a heterogeneous West African group. However, in the case of St. Vincent Island Black Caribs, there was considerable gene flow from the Creoles, which added European Gm haplotypes. Thus, the high level of genetic variation observed in the Garifuna, as estimated by the level of heterozygosity and the presence of genetic variants, is due to the particular patterns of admixture experienced by these populations.

Maintenance of genetic variation

The social organisation of the Black Caribs contributes to the maintenance of the observed level of genetic variation. In Livingston, Guatemala, approximately 45% of the Garifuna households are consanguineal and practise

Table 4: Serum protein gene frequencies in Black Carib populations

System and Allele	Belize		St. Vincent		Guatemala
	Stann Greek	Punta Gorda	Owia	Sandy Bay	Livingston
Albumin:					
Al^{A}	0.989	0.987	1.000	1.000	0.995
Al^{M}	0.011	0.013	0.000	0.000	0.005
Haptoglobin:					
Hp^{1}	0.531	0.561	0.439	0.685	0.596
Hp^{2}	0.446	0.439	0.433	0.315	0.394
Hp^{1M}	0.023	0.000	0.128	0.000	0.010
Transferrin:					
Tf^{B}	0.003	0.000	0.000	0.000	0.000
Tf^{C}	0.967	0.983	1.000	0.974	0.973
Tf^{D}	0.030	0.017	0.000	0.026	0.027
Group-specific component					
Gc^{1}	0.880	0.865	0.756	0.896	0.895
Gc^{2}	0.120	0.135	0.224	0.094	0.105
Gc^{Ab}	0.000	0.000	0.020	0.010	0.000
Properdin factor B (Bf):					
Bf^{B}	0.523	0.471	0.507	0.510	0.385
Bf^{F}	0.372	0.445	0.401	0.323	0.523
$Bf^{S0.7}$	0.002	0.000	0.086	0.153	0.004
Bf^{F1}	0.103	0.084	0.007	0.014	0.088

serial polygyny (Gonzalez, 1984). These households contain no marital pairs, but instead are composed of women and childrem assisted by brothers, uncles and adult sons. In such a household, the mother has children by a number of different fathers who establish resisdence for varying periods of time. When compared to permanent pair bonding, as observed in monogamous unions, serial polygyny recombines or "reshuffles" the genes within these populations at an accelerated rate. In addition to the genetic recombination, these mating patterns lower the likelihood of possible fetal wastage for any one woman through blood group incompatibilities.

The high reproductive rate in Black Carib populations also contributes to the level of genetic variation. Firschein (1961) observed a mean of 5.84 children in mothers over 60 years of age, while women in the same community but 45 to 54 years of age had produced 5.01 live births. These data suggest a recent small decline in the fertility of the Black Caribs. Brennan (1983) noted a similar fertility trend among the Honduras Garifuna. However the fertility levels were significantly higher in the rural, isolated villages with an observed mean of 10.9 live births per women 45 years of age or greater. Thus, in addition to the accelerated recombination rates of the genes due to serial polygyny, elevated fertility among the Black Caribs produces a large number of different genotypic combinations that are eventually tested in the diverse environments encountered during migration. Such a mating system is highly adaptive in a colonising population which produces small offshoots that migrate. Two related cultural mechanisms that contribute to the maintenance of the genetic variability in Black Carib populations are migration and hybridisation. Gene flow between the various ethnic groups provides a constant source of new genetic material into Black Carib communities. In 1972 in Livingston, Guatemala, 20 to 30% of all births had parents of differing ethnicities (Ghidinelli, 1976). Kerns (1984) observed a higher rate of interethnic matings in Black Carib towns when compared to rural villages. She estimated a total migration rate of approximately 15% between rural communities; 80% of these matings were between Black Caribs from different villages and only 20% were locally exogamous and between different ethnic groups. Based upon genetic data, Crawford (1983) observed that the coastal Belizean Black Carib communities receive gene flow from both the Creoles and the Maya Amerindians.

Genetic adaptation

Until recently, the Black Caribs of Central America have been under heavy malarial parasitisation. Custodio and Huntsman (1984) reported that from 1973 to 1978 in the Province of Colon, Honduras, malarial infestations were primarily due to *Plasmodium vivax* and *Plasmodium falciparum*, carried by the vectors *Anopheles albinamus* and *Anopheles darlingii*. Of the 80,822 persons tested for malaria, by the Society for the Eradication of Malaria in Honduras of 22.7% were positive for the *vivax* form and only 0.49% had been parasitised by *falciparum*.

Because of the West African genetic heritage, the Black Carib gene pool contains a number of genetic systems in which variants provide some protection against these malarial species. These systems include haemoglobin (with the genotypes AS and AC), glucose-6-phosphate dehydrogenase (A-genotype), the Duffy system with the FyFy genotype, and possibly the immunoglobulin allotypes, though to date a relationship between specific Gm haplotypes and malaria resistance has not been adequately demonstrated.

The Garifuna have been sampled more often for haemoglobinopathies than for any other genetic markers. Since Firschein's field research in British Honduras during 1956-7, a total of 3,969 Black Caribs have been screened for variant haemoglobins in five different projects. The incidence varies from 0.6% for the haemoglobin S allele in Sandy Bay (St. Vincent) to 23.6% in Seine Bight, British Honduras, in 1957 (Table 6). The overall Black Carib sickle cell carrier rate of 12.1% was computed from Custodio and Huntsman's data of 1972 and 1975, in which 1,750 persons were tested for abnormal hemoglobins. This rate is slightly higher than the 8 to 9% sickling prevalence observed in U.S. Black populations. However, this incidence of heterozygotes deviates significantly from the remarkably high HbAS frequency of 24% computed by Firschein (1961). The incidence of hemoglobin AC in the Black Caribs is lower than the expected frequencies, when their African ancestry is considered (Table 5).

Table 6 summarises the Hb^S and Hb^C gene frequencies for 12 populations of Black Caribs. Some of these groups have been resampled as many as four times over a period of 20 years. This resampling serves as a measures of the technical accuracy of the various laboratories and the adequancy of the sampling procedures, employed during the research. In Stann Creek there appears to be a gradient of Hb^S allelic frequencies from 11% to 8%, in surveys from 1956 to 1976, which it is tempting to attribute to the eradication of malaria in that area; however, a change in gene frequency of this magnitude is unlikely given a single generation, and on the basis of these sample sizes, sampling deviation is a more likely explanation for this variation. In Seine Bight, however, the observed difference in Hb^S, 23.6% in 1957 versus 11.6% in 1961, can be explained neither by selection relaxation nor by normal sampling variation, and the very high incidence of Hb^S observed in Seine Bight is probably the result of bias associated with the inclusion in the samples of at-risk mothers from a perinatal clinic.

The postulated selective mechanism for glucose-6-phosphate dehydrogenase deficiency acts through the malarial parasite's need for the oxidative pathway in red blood cell respiration. Thus, lowered levels of the enzyme are destructive to the parasite and reduce the likelihood of mortality from malaria. The incidence of Gd^{A-} (the deficient allele) is 9.7% in Hopkins and 7.8% in Stann

Table 5: Red blood cell protein gene frequencies in Black Carib populations

System and Allele	Belize		St. Vincent		Guatemala
	Stann Creek	Punta Gorda	Owia	Sandy Bay	Livingston
Haemoglobin:					
HbA	0.917	0.945	0.983	0.988	0.923
Hb^S	0.080	0.053	0.017	0.006	0.077
Hb^C	0.003	0.002	0.000	0.006	0.000
Esterase D:					
EsD^1	0.938	0.937	0.958	0.903	0.912
EsD^2	0.062	0.063	0.042	0.097	0.088
Acid phosphatase:					
p^A	0.183	0.146	0.148	0.141	0.175
p^B	0.804	0.836	0.852	0.853	0.783
p^C	0.000	0.002	0.000	0.000	0.000
p^D	0.000	0.000	0.000	0.000	0.000
p^R	0.013	0.017	0.000	0.006	0.042
Phosphoglucomutase - 1:					
PGM_1^1	0.838	0.850	0.847	0.829	0.879
PGM_1^2	0.162	0.150	0.153	0.171	0.121
6-Phosphogluconate dehydrogenase:					
PGD^A	0.978	0.960	0.997	0.989	0.983
PGD^C	0.022	0.040	0.003	0.011	0.017
Adenylate kinase:					
AK^1	0.973	0.971	1.000	0.997	0.988
AK^2	0.024	0.029	0.000	0.003	0.012
AK^3	0.003	0.000	0.000	0.000	0.000

Creek (Weymes & Gershowitz, 1984). The Duffy blood group system is involved in susceptibility and resistance to the *vivax* form of malaria. The FyFy genotype, which occurs at approximately 80% among the Black Caribs of Central America, prevents the parasite's entry into the erythrocytes and thus confers

Table 6: Haemoglobin variation among Black Carib populations

Population, date of fieldwork	Sample size	Allelic frequencies			References
		Hb^A	Hb^S	Hb^C	
Belize:					
Stann Creek, 1956	291	0.873	0.111	0.016	Firschein, 1961
Stann Creek, 1957	163	0.900	0.100	0.000	Firschein, 1961
Stann Creek, 1975	192	0.908	0.090	0.002	Weymes & Gershowitz, 1984
Stann Creek, 1976	349	0.917	0.080	0.003	Crawford et al, 1984
Hopkins, 1957	59	0.870	0.130	0.000	Firschein, 1961
Hopkins, 1974	257	0.969	0.030	0.001	Weymes & Gershowitz, 1984
Punta Gorda, 1976	237	0.945	0.053	0.002	Crawford et al, 1984
Punta Gorda, 1978	138	0.956	0.030	0.005	Weymes & Gershowitz, 1984
Seine Bight, 1956	147	0.872	0.116	0.012	Firschein, 1961
Seine Bight, 1957	64	0.718	0.236	0.046	Firschein, 1961
Seine Bight, 1975	127	0.915	0.075	0.010	Custodio & Huntsman, 1984
Honduras:					
Corozal, 1975	293	0.947	0.053	0.000	Custodio & Huntsman, 1984
Santa Roza de Aguan, 1972	457	0.902	0.090	0.008	Custodio & Huntsman, 1984
Tocamacho, 1975	209	0.955	0.045	0.000	Custodio & Huntsman, 1984
Punta Gorda, 1975	130	0.923	0.073	0.004	Custodio & Huntsman, 1984
Guatemala:					
Livingston, 1963	82	0.909	0.091	NT	Tejada et al, 1965
Livingston, 1975	202	0.923	0.077	0.000	Crawford et al, 1981
St. Vincent:					
Sandy Bay, 1979	163	0.988	0.006	0.006	Crawford et al, 1984
Owia, 1979	46	0.983	0.017	0.000	Crawford et al, 1984
Fancy, 1979	21	0.952	0.048	0.000	Crawford et al, 1984

total resistance against the *Plasmodium vivax* organism. This form of genetic adaptation has been termed "a cheap" adaptive process since there is no selective disadvantage associated with the FyFy genotype, such as occurs for example in the SS haemoglobin genotype.

The absence of abnormal haemoglobins, the FyFy genotype, and G6PD deficiency among indigenous Amerindians suggests that the plasmodium organism was introduced into the New World by infected Africans and/or Europeans from the Mediterranean region. Historical records document high malarial morbidity and mortality throughout the coastal settlements, decimating the Amerindian populations and subjecting the Black Caribs to strong selective pressures. Although the malarial vector has been eradicated in parts of Central America, morbidity from this disease has persisted into the 1970s in Honduras. Thus, their African heritage was a major factor in the evolutionary success of the Black Caribs and in their ability to colonise malarially infested regions of the New World.

GENETIC STRUCTURE

The allelic frequency-based affinities of the various subdivisions of the Black Carib population may be graphically represented by a least-squares approximation of a variance-covariance relationship matrix (Harpending & Jenkins, 1973). Figure 5 is a two-dimensional plot of eight Black Carib and Creole populations, according to the first two eigenvectors of the matrix based upon 14 genetic loci and 41 blood group and protein alleles. The Creole subdivisions were included in this analysis because of their proximal residence and the exchange of mates between the two ethnic groups. The first two scaled eigenvectors account for only 59.6% of the observed variation in the sample ($e_1\lambda^{\frac{1}{2}}$ = 35%, $e_2\lambda^{\frac{1}{2}}$ = 24.6%). While the addition of a third scaled eigenvector to a pseudo-three dimensional plot increases the total accounted variation to 74.8%, it is difficult to interpret the meaning of this plot (Crawford, 1983).

The subdivided population of St. Vincent shows greater genetic heterogeneity than do the coastal Central American Black Carib populations (Figure 5). Owia and Sandy Bay differ significantly from both Fancy (predominantly a Creole village) and the cluster of Central American Black Caribs and Creoles. This separation of Sandy Bay and Owia from the other groups reflects the high proportion of Amerindian ancestry. This is further evidenced by the contribution of haemoglobin A and the Rhesus chromosomal segments (cDE and CDe) to the dispersal of these communities along the first eigenvector (Figure 6).

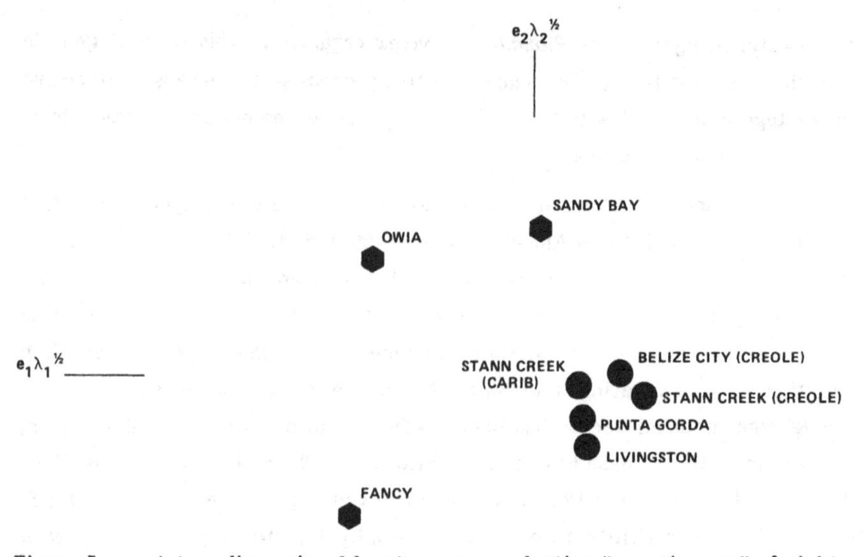

Figure 5: A two-dimensional least squares reduction "genetic map" of eight Black Carib and Creole populations based upon 41 allelic frequencies from 14 loci. St. Vincent Island populations are represented by hexagons, the coastal groups by circles. (From Devor et al, 1984.)

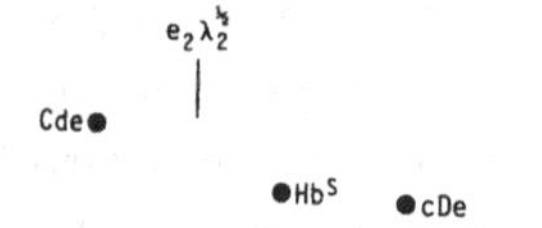

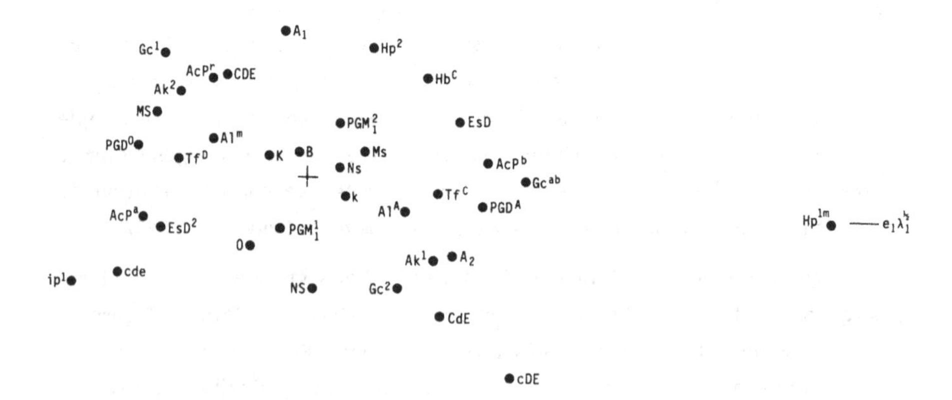

Figure 6: Least squares reduction of the 41 alleles and 14 loci corresponding to the "map" in Figure 4.

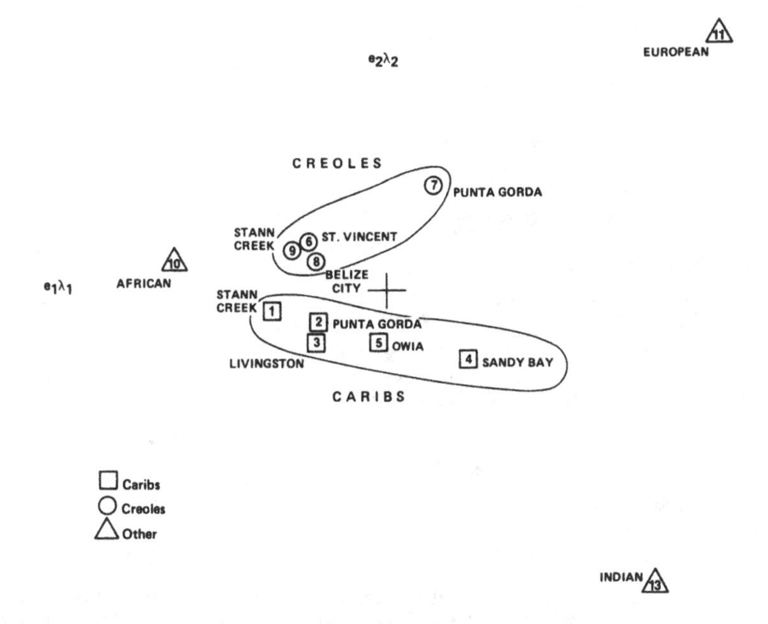

Figure 7: A least squares reduction "genetic map" of nine Black Carib and Creole populations compared to their parental groups, based upon immunoglobulin allotype frequencies. (From Schanfield et al, 1984.)

Similarly, Fancy, Livingston and Punta Gorda (all with considerable African ancestry) are separated from other groups, in part, by the elevated incidence of Hb^S and cDe. In addition, Hp1M and GcAb alleles, both relatively common in African groups, contribute to the separation of Fancy, Livingston and Punta Gorda from other populations.

The genetic structure analysis, based upon the frequencies of immunoglobulin allotypes, provides similar patterns of genetic affinity to these using traditional genetic markers. There is one major difference, the first two eigenvectors account for a greater proportion of the observed variation. Thus, a comparison of nine Black Carib population with three representative ancestral groups (West Africans, Europeans and Carib Indians) reveals that the first two eigenvectors subsume 93% of the variation. If the ancestral groups are deleted from the analysis, 83% of the variation is included. Apparently, the immunoglobulin markers provide finer discrimination between these triracial hybrid populations than do standard blood markers. The first eigenvector separates Sandy Bay and Owia from the Central American groups. Within the Creole cluster, Punta Gorda Creoles appear to be genetically distinct, but this is *probably* an artifact of the small sample size (Figure 7). The Black Carib

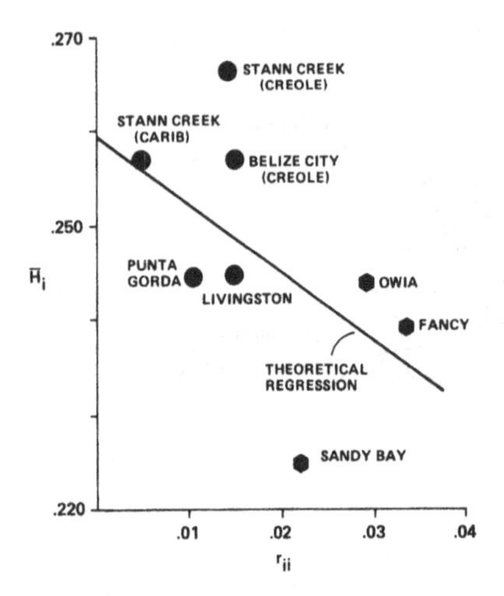

Figure 8: Plot of mean per locus heterozygosity (H_i) against distance from the centroid r_{ii} of the relationship matrix from eight Black Carib and Creole populations based upon frequencies for 41 alleles from 14 loci. St. Vincent Island populations are represented by hexagons, the coastal communities by circles. (From Devor et al, 1984.)

communities are clearly separated from the Creoles along the second eigenvector, with the Garifuna closest to the Indian parental group, while the Creoles show more affinity to the European group. The European gene flow is reflected in the Punta Gorda Creoles by a high incidence of the $Gm^{f;b}$ allotype. Sandy Bay is drawn toward the Amerindian parental representative because of the high incidence of $Gm^{zax;g}$, an Amerindian marker.

The genetic structure of the Black Carib populations is based upon the complex interaction between gene flow (admixture) on one hand, and the periodic fissioning of the gene pool, on the other hand. To disentangle the relative contributions of systematic (gene flow and selection) and nonsystematic (genetic drift) evolutionary pressures, the relationship between mean per locus heterozygosity (H_i) and genetic distances from the centroid of distribution (r_{ii}) were examined. The mean per locus heterozygosity based upon 41 alleles and 14 genetic loci is lower in Figure 8 because of the exclusion of the Gm locus. The insular nature of St. Vincent is reflected in the highest r_{ii} values in Fancy, Owia and Sandy Bay accompanied by a lower value of H_i. Stann Creek Creoles and

Table 7: Estimates of African versus Carawak admixture in Black Carib gene pools (Crawford et al, 1984)

Carib population	Percent African	Percent Carawak
Livingston	75	25
Stann Creek	67	33
Punta Gorda	70	30
Sandy Bay	46	54
Owia	61	39

Sandy Bay Caribs deviate most from the theoretical regression line, with the Creole group exhibiting the highest and Sandy Bay the lowest level of heterozygosity (probably resulting from different exposures to gene flow).

Admixture

The ethnohistory of the Black Caribs suggests that they are a biracial African/Amerindian hybrid population. Phenotypically the Black Caribs of Central America closely resemble West African Blacks. However, the Black Caribs of St. Vincent are less darkly pigmented and exhibit some Amerindian as well as African morphology. This, an initially biracial, hybrid model was examined using eight genetic loci (ABO, Rhesus, Duffy, MNSs, Diego, Kell, Kidd, and haemoglobin), 26 alleles and a true least squares method of analysis. Since the historical records support the existence of an Arawak/Carib hybridised group on St. Vincent, a "Carawak" parental group was constructed utilising an average of gene frequencies of Arawak and Venezuelan Caribs.

Table 7 summarises the estimates ($\underline{m}$) of the African and Carawak contribution to the Black Carib gene pool. Livingston has the highest proportion of African genes, Sandy Bay the lowest. These estimates suggest that the founders of Sandy Bay, who escaped deportation in 1797, were much more Indian than the Black Caribs who were deported. Owia contains intermediate Indian/African admixture because of high levels of Carib/Creole gene flow in that small village.

However on the bases of Gm allotype frequencies in which European haplotypes were detected in Black Caribs, Schanfield, Brown and Crawford (1984) estimated the proportion of African, European and Amerindian admixture in Black Carib and Creole populations using a modified Bernstein method (Table 8). The results show general agreement to those by standard blood markers and the addition of a European component usually reduces the Amer-

Table 8: Admixture estimates of Black Carib and Creole populations of Central America and the Caribbean (Schanfield et al, 1984)

	Admixture Percentages		
	African	European	Amerindian
Belize:			
Caribs			
Stann Creek	80	3	17
Punta Gorda	71	5	24
Hopkins	76	1	24
Creoles:			
Stann Creek	76	17	7
Belize City	75	17	8
Guatemala:			
Caribs			
Livingston	70	1	29
St. Vincent:			
Caribs			
Sandy Bay	41	17	42
Owia	58	10	32
Total Creoles	79	16	5

indian estimate. For example, in Sandy Bay the Carawak component of 54% was reduced to 42% in the Gm estimates of admixture. The high European admixture for Sandy Bay suggests gene flow from their Creole neighbours.

CONCLUSIONS

This study of the Black Caribs and Creoles of Central America and the Caribbean demonstrates once again the role of unique historical events in the genetic evolution of human populations and their resulting structure. The observed gene frequency distribution in this geographical area would make little evolutionary sense without cognizance of the patterns of migration and population fission. Similar relationships between history and genetics have been observed in numerous previous studies. However, in this Black Carib project, not only is history used to explain genetics but genetic data are utilised to resolve various historical ambiguities. For example, historians have been unable to agree upon the relative contributions of the Amerindians and West African slaves to the formation of the Black Carib population. The historical literature is replete with contradictory descriptions of the Indian or the African likeness of the Black Caribs. However the genetic estimates of admixture of Central American Black Caribs paint a consistent picture. The genetic results also document gene flow from the Highland Maya to the Black Carib and Creole gene pools, despite historical claims of reproductive isolation.

Cavalli-Sforza (1983) notes "...it is more interesting if reasonable models do not fit than if they do, provided one actively looks for the reasons for the discrepancy. There can be something of interest in the analysis of exceptions..., especially if one can identify possible reasons." This study of the Black Caribs has indeed produced such an exception. The results of earlier studies of Black Caribs from Livingston revealed that 70% of the gene pool was of West African origin and 30% was Amerindian (Crawford et al, 1981). Similar estimates were obtained for other Central American Garifuna populations, with the African component varying between 67% and 80%. It was surprising to discover that the Black Carib gene pool on St. Vincent Island, the founding group for all of the coastal populations, contained only 41% African genes. This discrepancy can be explained by a series of hypotheses, as follows: (1) The 2,026 Black Caribs who were deported from St. Vincent were more Amerindian than the 30% estimated in the contemporary population, but, as they colonised the coast of Central America, West African enclaves were incorporated. (2) The more "Indian" individuals escaped deportation and founded the Black Carib population of St. Vincent Island. Those of African ancestry were shipped to Roatan. (3) One possible mechanism, that can explain the disproportionate African ancestry of those shipped to Roatan, may involve the "typhus" epidemic on Balliceaux Island. Gonzalez (1984) estimated that 4,200 Black Caribs were imprisoned by the British on Balliceaux from the end of the war (July, 1796) to the arrival of the transport (March, 1797). During this period, a total of 2,174 Black Caribs died from an epidemic. If the New World populations were more susceptible to this disease, then selection may have been responsible for the composition of the deported population. (4) Natural selection may have favored those individuals who were resistant to malaria and thus had a more African genotype. Apparently malaria had become endemic on the coast of Honduras, and individuals with HbAS, HbAC, FyFy and G6PD deficiency would have had an advantage. This "discrepancy" based upon historical models offers the possibility of testing important evolutionary questions.

Finally, the population dynamics of colonizing tropical populations can be understood and modelled on the basis of the Black Caribs. Their evolutionary success is in part due to high levels of fertility, elevated genetic variation (due to their triracial origins and their unique social organisation) and their adaptation to malaria. Recently, the Black Caribs have ceased their relentless geographical expansion along the coast of Central America and are developing colonies in large cities such as New York. It will be interesting to monitor the demographic transition of this tropical population in an urban environment.

ACKNOWLEDGMENTS

I would like to thank Lori Houk-Stephens for typing this manuscript and William Davison for the use of several illustrations.

REFERENCES

Brennan, E. R. (1983). Factors underlying decreasing fertility among the Garifuna of Honduras. American Journal of Physical Anthropology, **60**, 177.

Crawford, M. H. (1983). The anthropological genetics of the Black Caribs of Central America and the Caribbean. Yearbook of Physical Anthropology, **25**, 155-186.

Crawford, M. H., Dykes, D. D., Skradsky, K. & Polesky, H. F. (1984). Blood group, serum protein, and red cell enzyme polymorphisms, and admixture among the Black Caribs and Creoles of Central America and the Caribbean. In: M.H. Crawford (ed.), Current Developments in Anthropological Genetics, Volume III. Black Caribs: A Case Study of Biocultural Adaptation, pp. 303-333. New York: Plenum Press.

Crawford, M. H., Gonzalez, N. L., Schanfield, M. S., Dykes, D. D., Skradsky, K. & Polesky, H. F. (1981). The Black Caribs (Garifuna) of Livingston, Guatemala. Human Biology, **53**, 87-103.

Crawford, M. H., Workman, P. L., McLean, C. & Lees, F. C. (1976). Admixture estimates and selection in Tlaxcala. In: M.H. Crawford (ed.), The Tlaxcaltecans: Ethnohistory, Demography, Morphology and Genetics, pp. 161-169. Lawrence: University of Kansas Anthropology Series.

Custodio, R. & Huntsman, R. (1984). Abnormal hemoglobins among the Black Caribs. In: M. H. Crawford (ed.), Current Developments in Anthropological genetics, Volume III. A Case Study of Biocultural Adaptation, pp. 335-343. New York: Plenum Press.

Davidson, W. V. (1984). The Garifuna in Central America: Ethnohistorical and geographical foundations. In: M.H. Crawford (ed.), Current Developments in Anthropological Genetics, Volume III. Black Caribs: A Case Study of Biocultural Adaptation, pp. 13-35. New York: Plenum Press.

Devor, E. J., Crawford, M. H. & Bach-Enciso, V. (1984). Genetic population structure of the Black Caribs and Creoles. In: M.H. Crawford (ed.), Current Developments in Anthropological Genetics, Volume III. Black Caribs: A Case Study of Biocultural Adaptation, pp. 303-333. New York: Plenum Press.

Firschein, I. L. (1961). Population dynamics of the sickle-cell trait in the Black Caribs of British Honduras, Central America. American Journal of Human Genetics, **13**, 233-254.

Ghidinelli, A. (1976). La familia entre los Caribes Negros, Ladinos y Kekchies de Livingston. Guatemala Indigena, **11**, 1-315.

Gonzalez, N. (1984). Garifuna (Black Carib) Social Organization. In: M.H. Crawford (ed.), Current Developments in Anthropological Genetics, Volume III. Black Caribs: A Case Study of Biocultural Adaptation, pp. 51-65. New York: Plenum Press.

Gullick, C. J. M. R. (1976). Exiled from St. Vincent: The Development of Black Carib Culture in Central America Up to 1945. London: Progress Press.

Gullick, C. J. M. R. (1979). Ethnic interaction and Carib language. Journal of Belizean Affairs, **9**, 3-20.

Gullick, C. J. M. R. (1984). The changing Vincentian Carib population. In: M.H. Crawford (ed.), Current Developments in Anthropological Genetics, Volume III. Black Caribs: A Case Study of Biocultural Adaptation, pp. 37-50. New York: Plenum Press.

Harpending, H. C. & Chasko, W. J. (1976). Heterozygosity and population structure in Southern Africa. In: E. Giles & J.S. Freidlaender (eds.), The Measures of Man: methodologies in Biological Anthropology, pp. 214-229. Cambridge: Peabody Museum Press.

Harpending, H. & Jenkins, T. (1973). Genetic distances among Southern African populations. In: M.H. Crawford & P.L. Workman (eds.), Methods and Theories of Anthropological Genetics, pp. 177-199.

Harpending, H. C. & Ward, R. (1982). Chemical systematics and human populations. In: M. Nitecki (ed.), Biochemical aspects of Evolutionary Biology, pp. 213-256. Chicago: University Chicago Press.

Kerns, V. (1984). Past and present evidence of interethnic mating. In: M.H. Crawford (ed.), Current Developments in Anthropological Genetics, Volume III. Black Caribs: A Case Study of Biocultural Adaptation, pp. 95-114. New York: Plenum Press.

Krieger, H., Morton, N. E., Mi, M. P., Ezevedo, E., Friere-Maia, E. & Yasuda, N. (1961). Racial admixture in north-eastern Brazil. Annals of Human Genetics, **29**, 113-125.

Reed, T. E. & Schull, W. J. (1968). A general maximum likelihood estimation program. American Journal of Human Genetics, **20**, 579-580.

Roberts, D. F. & Hiorns, R. W. (1965). Methods of analysis of genetic composition of a hybrid population. Human Biology, **37**, 38-43.

Rouse, I. (1976). Cultural development and Antigua, West Indies: A progress report. Actas XLI Congress International Americanistas, **3**, 701-709.

Schanfield, M. S., Brown, R. & Crawford, M. H. (1984). Immunoglobulin allotypes in the Black Caribs and Creoles of Belize and St. Vincent. In: M.H. Crawford (ed.), Current Developments in Anthropological Genetics, Volume III. Black Caribs: A Case Study of Biocultural Adaptation, pp. 345-363. New York: Plenum Press.

Taylor, D. (1951). The Black Caribs of British Honduras, Viking Fund Publications in Anthropology. No. 17. New York: Wenner-Gren Foundation.

Tejada, C., de Gonzalez, N. L. S. & Sanchez, M. (1965). El Factor Diego y el gene de cleulas falciformes entre los Caribes de raza Negra de Livingston,

Guatemala. Reimpreso Revista Collegio Medicina, Guatemala, **16**, 83-86.

Weymes, H. & Gershowitz, H. (1984). Genetic structure of the Garifuna population in Brasil. In: M.H. Crawford (ed.), Current Developments in Anthropological Genetics, Volume III. Black Caribs: A Case Study of Biocultural Adaptation, pp. 271-287. New York: Plenum Press.

HISTORICAL AND DEMOGRAPHIC FACTORS AND THE GENETIC STRUCTURE OF AN AFROAMERICAN COMMUNITY OF NICARAGUA

G. BATTISTUZZI[1], G. BIONDI[2], O. RICKARDS[3],
P. ASTOLFI[4], G. F. DE STEFANO[3]

[1]*Dipartimento di Genetica, Biologia Generale e Molecolare, Universita di Napoli, Napoli, Italy*
[2]*Dipartimento di Biologia Animale e dell'Uomo, Universita di Roma "La Sapienza", Rome, Italy*
[3]*Dipartimento di Biologia, Universita di Roma "Tor Vergata", Rome, Italy*
[4]*Dipartimento di Genetica e Microbiologia "Adriano Buzzati-Traverso", Universita di Pavia, Pavia, Italy.*

INTRODUCTION

Peopling of uninhabited regions, migration, intermixture and isolation have certainly played a major role in the evolutionary history of human populations. The communities of Afroamerican ancestry spread along the Atlantic coast of Central America and on the Caribbean Islands provide an opportunity to evaluate the interplay of such phenomena in the natural history of a human population and their relative importance in the determination of its genetic structure.

These Afroamerican communities are thought to be the result of a complex process of admixture involving, to an extent which varies from one community to another, various African peoples, mostly from central and western Africa, Europeans and American Indians (Firschein, 1961; Ryan et al, 1970; Benoist, 1971; Salzano, 1971; De Stefano, 1979; De Stefano et al, 1982; Crawford, 1983 a,b; De Stefano & Calicchia, 1985). The present paper reports the results of the analysis of historical records, demography and genetic structure of the Afroamerican population inhabiting the Atlantic coast of Nicaragua, in an attempt to reconstruct the patterns that have led to the present-day population and to estimate the relative ethnic contributions to its gene pool.

EARLY HISTORY

Before the contact with Europeans the territory of Nicaragua was settled by Indian populations whose origin and interrelationship are not yet fully understood (Figure 1). The Chorotega and Subtiaba Indians, very likely related

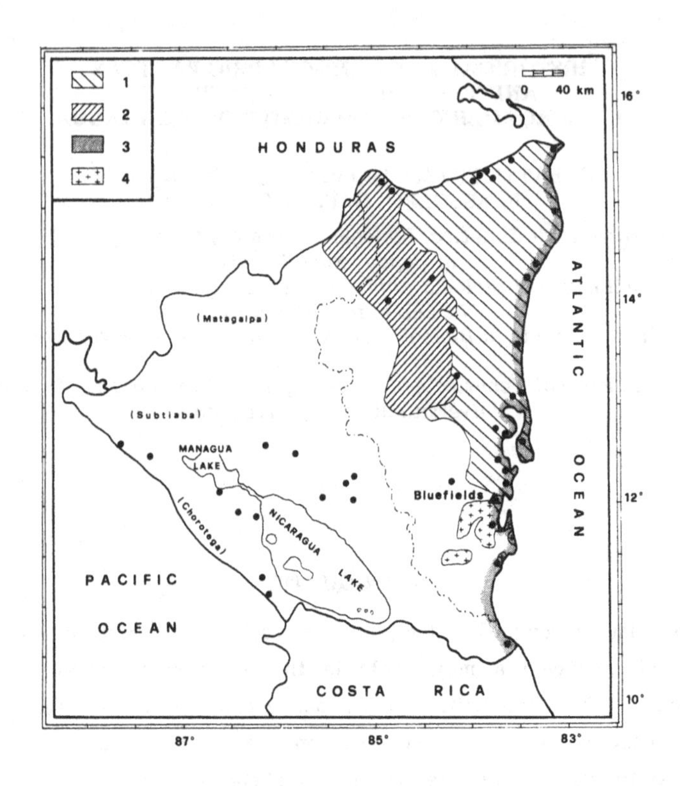

Figure 1: Map of Nicaragua showing location of the major ethnic groups. Numbers indicate: (1) Miskito, (2) Sumo, (3) Afroamerican, and (4) Rama settlements. Ladinos inhabit the western regions of Nicaragua. The eastern area delimited by the dashed line corresponds to the Department of Zelaya. Full circles indicate the places of birth of the subjects included in the sample for the analysis of migration.

to the Indians of Central and North America (Thomas & Swanton, 1911; Sapir, 1925; Biasutti, 1967; Nietschmann, 1969; Ryan et al, 1970; De Stefano, 1971; Massajoli & De Stefano, 1981 a,b), peopled the Pacific coastal regions of Nicaragua. The Matagalpa Indians inhabited the western-central regions. Their origin and relationship with other Indian populations of Nicaragua are still disputed, though they are likely to be related to the Indians of South America (Adams, 1957; Johnson, 1962; Kidder, 1962; De Stefano, 1971). Eastern regions were peopled by hunting and fishing Indian populations of Chibcha language definitely related to the South American Indians (Massajoli, 1968; Nietschmann, 1969; De Stefano, 1971; Massajoli & De Stefano, 1981 a,b). The Sumo Indians, living along the river valleys of central Zelaya, were and are the most isolated group. The Miskito and Rama Indians were located mostly along the Atlantic coast north and south of the Bluefields lagoon respectively. Due to the various processes of intermixture that have occurred during the last four centuries, very small remnants, if any, of these original populations are present today, in spite of the existence of several groups who still maintain their ethnic definition and language.

European domination over the country caused a dramatic diversifying effect on the population of Nicaragua. The Spanish, whose domination began in 1522, limited their influence to the western populations, massively intermarrying with and causing the fusion of the autochthonous populations (Rosenblatt, 1954). As a result of this process, a new hybrid population arose, the Ladino population, which accounts for 70-80% of the present-day population of Nicaragua. The British, whose domination started two centuries later and was restricted to the eastern regions of Nicaragua and more precisely to the coastal areas, adopted a different colonising strategy, creating new human settlements of people imported from Africa and discouraging their admixture with the local Indian populations.

The history of the Afroamerican population of Zelaya can be traced, according to Ryan et al (1970), to a small number of survivors of a shipwreck at Cabo Gracias a Dios, at the extreme north of the Atlantic coast of the country. However, it was only at the beginning of the British rule (1695-1850) that permanent settlements were established (Figure 1). These settlements, all along the Coast of Mosquitia (east coast), were mostly the result of importation of slaves from western and central Africa. Their number and strength was considerably increased during the 18th and 19th centuries by runaway slaves from European-held regions, such as the Caribbean Islands, United States and South America, by slaves deported from the Islands, and by people migrating from the neighbouring Afroamerican communities of Central America such as the Black Caribs during the 19th century (Crawford, 1983 a,b). Thus, importation of Africans, the major component of early immigration to Mosquitia, progressively subsided and was replaced by immigration of people of African ancestry already intermixed with American Indians or Europeans or both. With the growth of the communities the relative amounts of migration to and from Zelaya changed and, during this century, have become virtually equal. Only during the 1920s was there net immigration, connected with the establishment of banana plantations.

RECENT HISTORY

Since the time of their first settlements and although strictly prohibited by British law, the incomers intermarried relatively freely with Miskito and, more recently, with Rama Indians. By contrast, in the history of the population of Zelaya the active avoidance of intermixture and integration between Western, Ladino, and Eastern populations is quite well documented. There were continual movements towards autonomy, e.g. the foundation of an independent Miskito kingdom (1880-1894) and, more recently (1927-1933), the Sandino rising (Massajoli & De Stefano, 1981 a,b). In order to achieve better integration of Western and Eastern populations, the central Government of Nicaragua promoted, in 1960, a major campaign of internal migration towards the east coast. The demographic success of the campaign is shown by the increased proportion and spread of the Ladinos in Zelaya. In 1960 they accounted for 70% of the total population of Nicaragua and made up 15% of the population of Zelaya where they were concentrated in the central mining centres. Ten years later they accounted for 25% of the Zelaya population and were found in all the major towns of the province.

Census data show that 75% of the non-Ladino population of Zelaya is composed of Indians (20%) and Afroamericans (55%). These estimates are based on criteria of classification which include a variety of characteristics such as self definition, language, religion, and place of birth. However, there is only a loose, if indeed any, correlation of self definition with appearance, anthropometric and other biological characteristics. Moreover a person with some African ancestry may use a variety of self definitions such as Black, Mulatto, Creole, Zambo, which are not always synonymous. For instance, Creoles form a fairly endogamous group of English culture, Moravian religion and with usually a higher than average educational level. The Zambos, on the other hand, are a rather variable group who speak Spanish, by contrast to the other groups of African ancestry. This group certainly shows that to a certain extent cultural, and possible physical, intermixture between East and West has occurred (Massajoli & De Stefano, 1981 a,b).

MIGRATION AND MATING

A more detailed analysis of the patterns of migration and mating which characterises the last 70 years of the present century of the population of African ancestry of Nicaragua has been carried out on a selected sample of the population of Bluefields (Figure 1), one of the major and more representative

Table 1: Migration rates according to the generation and the area of origin of the subjects

Place of birth	Propositi %	Parents %	Grandparents %
Bluefields	70.8	77.2	77.5
Zelaya:			
Atlantic coast	15.7	11.6	6.3
Northern-central	8.1	1.7	1.9
Western Nicaragua	2.7	3.7	3.6
Antilles	1.1	2.1	6.0
Other countries	1.6	3.7	4.7
Crude internal migration rate	26.5	17.0	11.8
Crude total migration rate	29.5	22.8	22.5
m_e	25.0	18.9	19.6
Endogamy of Bluefields		42.7	40.6

towns of the west coast. Very high total immigration, ranging from 22.5% in the grandparental generation to 29.5% in the propositi, characterises the population (Table 1). Migration from abroad (Antilles and other countries) dramatically decreases from approximately 50% of the total migration in grandparents to about 10% in propositi. The higher figures observed in grandparents and parents are almost certainly a result of the establishment of the banana plantations in the two decades 1920-1940. However, since this analysis refers only to the portion of immigrants that left progeny in Bluefields and since the total amount of migration remains relatively constant or even increases over the three generations, the drop in migration from abroad is possibly an index of changing criteria of mate selection, mates from Zelaya being preferred to those from abroad. Internal migration is largely of people from Zelaya, while the contribution to total migration by people from the Ladino area is relatively low and constant. The migrant classification was based on place of birth, not on ethnic definition, but even if the latter contribution entirely represents those derived from intermixture it seems clear that marriages between Ladinos and Afroamericans are few and possibly deliberately avoided, and that the process of integration desired by the Central Government, whose effects should have appeared in the parental and propositus generations, is not taking place. On the other hand the increase of migration from north and central Zelaya, areas peopled mostly by Indian populations, possibly means an increased rate of intermixture between Indian and Afroamerican populations. Such an increase is to be viewed both as an increase of traditional intermixture

Table 2: Mean value of distances of migration from Bluefields for the three generations under study

	Atlantic coast (Afroamerican area)	Zelaya (Indian area)	Western Nicaragua (Ladino area)	Antilles
Propositi	107.4±15.3	217.6±14.6	183.1±51.2	1440.0
Parents	90.5± 8.1	212.1±19.6	197.8±27.2	1188.0±76.8
Grandparents	88.5±10.1	226.4±12.9	219.6±12.4	1022.4±28.5

Distances are expressed in kilometers. Mean values are the weighted mean of the distances between places of birth grouped according to the area.

between Afroamericans and Miskito and of the extension of the admixture process to Rama and, indirectly, to Sumo Indians (De Stefano, 1971).

The quantitative variation here observed in the rates of migration are not associated with qualitative changes. In fact the number and location of birth places within each of the areas mentioned has remained constant, over the three generations, as shown by the absence of significant variation of mean distance between places of birth and Bluefields (Table 2). The distribution of the places of birth of the subjects in the three generations examined shows a good correlation with the geographical distribution of the major communications, the roads along which most of the trade activity involving the population of Bluefields is carried on. Finally, as far as internal migration is concerned, sampling within each area is proportional to the size of the population and seems not to be correlated with distances from Bluefields.

The analysis of the historical, cultural and demographic events influencing the natural history of the Afroamerican population of Nicaragua suggests a close relationship between this population and the neighbouring Afroamerican communities. Moreover these data strongly support the hypothesis of its trihybrid origin, West African and Indian populations having contributed the major portion of the gene pool, while the European contribution is relatively small. An estimate of how close the population of Bluefields is to other Afroamerican communities and of how much the three components of the hybrid population have contributed to it has been attempted by the analysis of the gene frequencies of a number of blood group, red blood cell enzyme and protein, and plasma protein systems. The one dimensional eigenvectorial representation of an R matrix based upon red cell and plasma protein polymorphisms (Figure 2) gives a fairly clear idea of the amount of European gene flow in Afroamerican

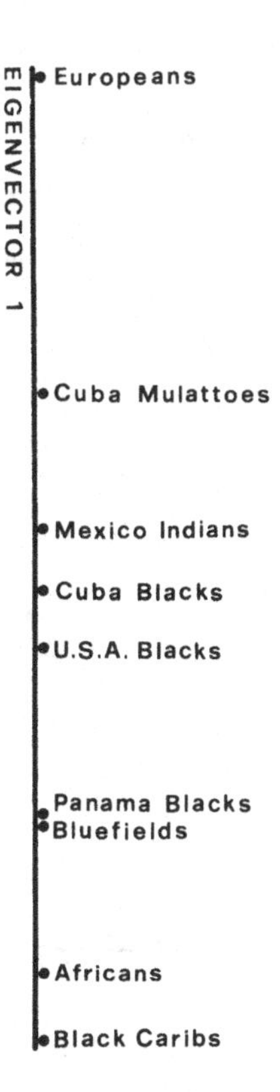

Figure 2: Relative contribution of West African and European populations to the gene pool of various Afroamerican communities. A one-dimensional eigenvectorial representation of an R matrix based upon the frequencies of PGM, PGD, AcP and Hp polymorphisms (Biondi et al, 1985). The first principal component, here represented, accounts for 63%of total variance of gene frequencies.

communities. The populations of Bluefields, of Panama and the Black Caribs are clustered close to the West African. European gene flow is greater for the populations of USA and Cuban Blacks and Cuban Mulattos. A similar analysis based upon the gene frequencies of blood groups is shown in Figure 3. The second principal component successfully discriminates among the Afroamerican populations those in which the gene flow has been predominantly from Europeans and those in which the gene flow is predominantly from Indian populations, and parallels the observations of Crawford (1983 a,b) on the communities of Black Caribs of Central America. The population of Bluefields is located, together with the Black Caribs of Livingstone (L) and of Sandy Bay (SB), along the axis

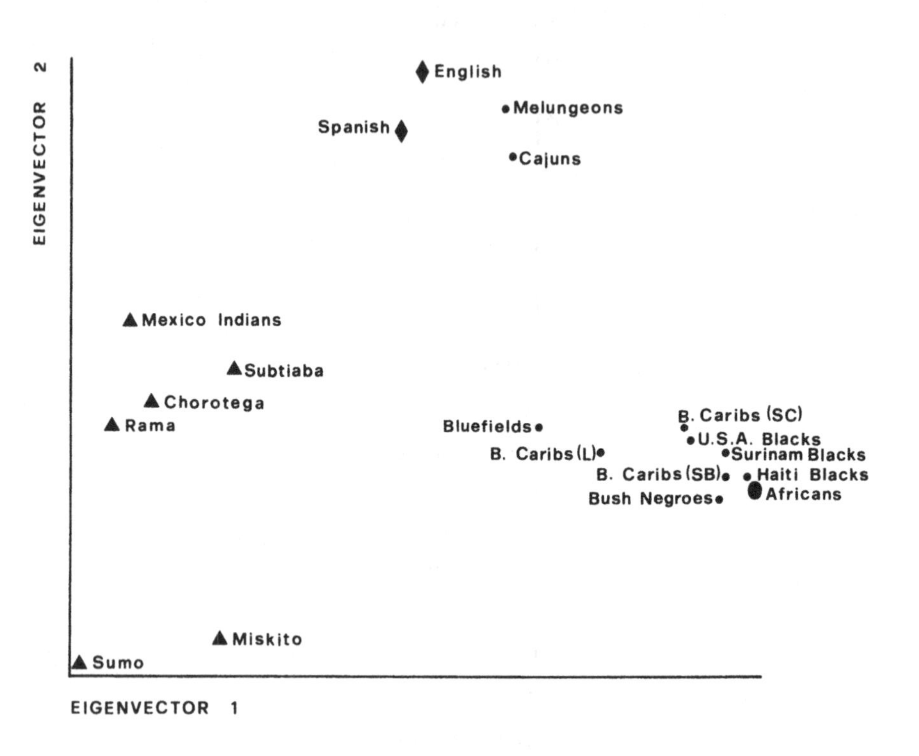

Figure 3: Gene flow from European or Indian populations differentiate Afroamerican communities.
A two-dimensional eigenvectorial representation of an R matrix based upon the frequencies of ABO, Rh and MN blood groups (Biondi et al, 1983 and unpublished data). The first and second principal components account for 87% of total variance of gene frequencies.

connecting the centroids of the American Indian and West African populations, meaning that the admixture process which characterises these populations has involved mostly Africans and American Indians. These results can be quantitatively expressed in terms of relative contributions to the gene pool of the hybrid population (Table 3). They show that the Afroamerican population of Nicaragua can be considered as a hybrid population in which about two thirds of the gene pool has been contributed by West African populations and one third by Indians. The European contribution is very small, and the estimate does not significantly differ from zero.

Table 3: An estimate of African, American Indian and European admixture in the Afroamerican population of Nicaragua.

Populations	Contribution (%)
West Central Africans	66
Europeans	1
Indians	33

Admixture was estimated by least squares method (Elston, 1971). Estimates were calculated for the population of Bluefields, which we have shown to be representative of the Afroamerican population of Nicaragua (Biondi et al, 1983, and unpublished data).

CONCLUSION

This analysis of the historical, cultural and demographic events affecting the natural history of the Afroamerican population of the Atlantic coast of Nicaragua suggests a close relationship between this population and the other Afroamerican communities. The genetic data support the hypothesis of its trihybrid origin, West African and Indian populations having contributed the major portion of the gene pool, the African about two thirds and the Indian one third, while the European contribution is relatively small.

REFERENCES

Adams, R. N. (1957). Cultural Surveys of Panama-Nicaragua-Guatemala-El Salvador-Honduras. Pan American Sanitary Bureau, Scientific Publications, **33**, Washington.

Benoist, J. (1971). Population Structure in the Caribbean Area. In: F.M. Salzano (ed.), The Ongoing Evolution of Latin American Populations, pp. 221-249. Springfield: Charles C. Thomas.

Biasutti, R. (1967). Le Razze e i Popoli della Terra. 4 ed. Torino: UTET.

Crawford, M. H. (1983a). The Anthropological Genetics of the Black Caribs (Garifuna) of Central America and the Caribbean. Yearbook of Physical Anthropology, **26**, 161-192.

Crawford, M. H. (ed.) (1983b). Current Developments in Anthropological Genetics. Vol. 3. The Black Caribs: a Case Study of Biocultural Adaptation. New York: Plenum.

De Stefano, G. F. (1971). Richerche di Antropologia Biologica su Popolazioni Nicaraguensi. Rivista di Antropologia, **57**, 205-220.

De Stefano, G. F. (1979). Sensibilita alla F.T.C. e Gruppi Sanguigni (ABO, Rh-D) in un campione di Creols di Bluefields, Nicaragua. Rivista di Antropologia, **60**, 99-104.

De Stefano, G. F., Biondi, G. & Battistuzzi, G. (1982). Population Intermixture: a Multidisciplinary Study on the Population of Bluefields, Nicaragua. II Congress of the European Anthropological Association, Brno, 1980. Anthropos, **22**, 129-132.

De Stefano, G. F. & Calicchia, M. C. (1985). Digital Dermatoglyphics in two Afroamerican communities of Central America. Antropologia Contemporanea (in press).

Elston, R. C. (1971). The Estimation of Racial Admixture in Racial Hybrids. Annals of Human Genetics, **35**, 9-17.

Firschein, I. L. (1961). Population Dynamics of Sickle-Cell Trait in Balck Caribs of British Honduras, Central America. American Journal of Human Genetics, **13**, 233-254.

Johnson, F. (1940). The Linguistic Map of Mexico and Central merica. In: C.L. Hay et al (eds.), The Maya and Their Neighbors, pp. 88-114. New York: Appleton-Century Co.

Kidder, A. (1940). South American Penetrations in Middle America. In: C.L. Hay et al (eds.), The Maya and Their Neighbors, pp. 441-459. New York: Appleton-Century Co.

Massajoli, P. L. (1968). I Sumo e i Miskito. L'Universo, **48**, 727-780.

Massajoli, P. L. & De Stefano, G. F. (1981a). La Regione Orientale del Nicaragua. I and II. L'Universo, **61**, 9-46.

Massajoli, P. L. & De Stefano, G. F. (1981b). La Regione Orientale del Nicaragua. III. L'Universo, **61**, 257-278.

Nietschmann, B. (1969). The Distribution of Miskito, Sumo and Rama Indians, Eastern Niacaragua. Bulletin of the International Committee on Urgent Anthropological and Ethnological Reesearch, **11**, 91-101, Vienna.

Rosenblat, A. (1954) La Poblacion Indigena y el Mestizaje en America, vols. 1 and 2. Editorial Nova. Buenos Aires.

Ryan, J. M., Anderson, R. N., Bradley, H. R., Negus, C. E., Johnson, R. B., Hanks, C. W. Jr., Crotean, G. F. & Council, C. C. (1970). Area handbook for Nicaragua. New York: Johnson Research Associates.

Salzano, F. M. (1971). Retrospect and Prospect. In: F.M. Salzano (ed.), The Ongoing Evolution of Latin America Populations, pp. 679-691. Springfields: Charles C. Thomas.

Sapir, E. (1925). The Hokan Affinity of Subtiava in Nicaragua. American Anthropology, **27**, 402-434.

Thomas, C. & Swanton, J. R. (1911). Indian Languages of Mexico and Central America and Their Geographical Distribution. Bulletin of the Smithsonian Institution Bureau of American Ethnology, **44**, Washington.

MIGRATION AND GENETIC POLYMORPHISMS IN SOME CONGO PEOPLES

G. SPEDINI[1], F. MENCHICCHI[1], G. DESTRO-BISOL[1] and M. SCHANFIELD[2]

[1]*Dipartimento di Biologia Animale e dell'Uomo, Cattedra di Antropologia, Universita "La Sapienza" Rome, Italy.*
[2]*Genetic Testing Institute, Atlanta, Georgia, U.S.A.*

INTRODUCTION

In ancient times, the equatorial forest and the Sahara desert were veritable geographical barriers to the migration of human populations from the Sudan to areas elsewhere in Africa. About the year 300 BC, during the Nok civilisation of the Niger-Cameroons plateau the central and southern regions of the African continent began to be colonised by the Proto-Bantu. These peoples had learned how to work iron and so were helped in this expansion by their newly developed stronger tools to clear ways through the forest and thus came to settle in the open territories to the south (Ki Zerbo, 1977; Alexandre, 1981). From this slow but irresistible flow of these early groups, all the Bantu populations settled in the African regions south of the northern hinge of the rain mantle are thought to be derived.

The present Bakongo and Bateke populations of southern Congo, the Pool region and the central plateau (Cuvette region), together with the Beti of its more northern regions, are among those whose ancestors participated in these migrations, though at different historical times. The powerful Mokoko Teke kingdom was already established in southern Congo by the fifteenth century AD and independent of the pre-existing Kongo kingdoms (Oliver & Fage, 1974; Ki Zerbo, 1977; Alexandre, 1981; Mair, 1981).

The migration to the north-west Congo was much later. It was only in the eighteenth century that Beti populations of the Mbochi ethnic group settled along the northern border of the Congo forest, forced south from the Sudan savannah lands by pressure from invaders said to be "centaures rouges au corps couvert d'ecailles" (Alexandre, 1981). As evidence of their recent separation from the Sudan peoples, the Beti still retain a few cultural elements, including an exogamous patrilineal social organisation.

The Babenga, a contraction of the original term for pygmies of Congo and Obangui, the Ba-Mpenga or "the men with a spear", are found all over the territory of Congo, particularly in the Pool region. They are regarded as indigenous as their history involves the earlier settlements of pygmies in the rain forest. While a few Babenga groups still hunt and maintain some of the original distinguishing characters, the western Pool Babenga have become sedentary, and have to a large extent lost their original identity.

The population of Congo was 1,537,000 in 1980. The Bateke represented around 22% of the total population, and the Beti about 13%. The Babenga in the Pool region were estimated at 272 in 1955 (Vallois & Marquer, 1980).

MATERIAL AND METHODS

During an investigation by Professor Cresta in two villages of the northern Congo (Cuvette region) and two in the southern Congo (Pool region), blood specimens were obtained by venepuncture. These were refrigerated and taken to Brazzaville, and from there to the Anthropology Laboratory in Rome. The specimens reported in this investigation were from 95 Beti, 80 Bateke of the Cuvette region, 57 Bateke and 47 Babenga of the Pool region. The division into groups was by ethnic affiliation and, for the Bateke, by geography. Biochemical markers were examined using both the conventional electrophoretic techniques on cellulose acetate (Chemetron, Labometrics s.r.l., Milan) and isoelectric focusing on ultra thin layer polyacrylamide gel as practised in the Hemotypology Laboratory at the Institute for Forensic Medicine of the Sacred Heart Catholic University in Rome. In addition, Km immunoglobulins were typed at the Genetic Testing Institute of Atlanta, Georgia.

The following erythrocyte and serum polymorphisms were analysed: acid phosphatase (Hopkinson et al, 1963), adenylate kinase (Fildes & Harris, 1966), glyoxalase I (Kompf et al, 1975), beta-hemoglobin (Golias, 1971), 6-phosphogluconate dehydrogenase (Fildes & Parr, 1963), phosphoglucomutase (Spencer et al, 1964; Bark et al, 1976), superoxide dismutase (Meera Khan, 1971), ceruloplasmins (Martin et al, 1961) and haptoglobins (Smithies, 1955).

RESULTS AND DISCUSSION

The distribution of phenotypes and the corresponding gene frequencies are reported in Tables 1 and 2 respectively. The differences between the observed and expected frequencies are not statistically significant with the exceptions of the Km system in southern Bateke ($\chi^2_1 = 4.60$, $p < 0.05$), and of PGM_1 in the Beti ($\chi^2_1 = 4.67$, $p < 0.05$). On the whole, the gene frequencies in the four populations fall within the ranges for sub-saharan Africa (acid phosphatase: P^a from .046 to .344; P^r from .000 to .233; 6-phosphogluconate dehydrogenase: PGD^c from .000 to .102; glyoxalase I: GLO^1 from .175 to .391; beta-haemoglobin: β^S from .003 to .201; phosphoglucomutase locus 1: PGM_1^1 from .702 to .970 (Jenkins & Corfield, 1972; Jenkins & Nurse, 1974; Hiernaux, 1976; Bender et al, 1977; Nurse et al, 1979; Vergnes et al, 1979; Santachiara-Benerecetti et al, 1980; Hitzeroth et al, 1981; Nurse & Jenkins, 1982; Le Gall et al, 1982)).

It is interesting that two variants of 6-PGD occurred in these samples. The first in a Babenga is presumably of the Richmond type described by Carter et al (1968) in an American negro. This phenotype is quite rare; absent among the Europeans and Mongoloids it occurs at low frequencies in Africa, mainly among Bantus, but is not described so far in Pygmies. The second variant, found in a northern Bateke, is quite different from all those reported so far in the literature.

Of some interest also is the low frequency of the acid phosphatase allele P^r, the "negro allele". This is in line with the findings in other areas of highly endemic malaria (Spedini et al, 1980), where the low frequencies were tentatively attributed to a possible disadvantage of the P^r allele in such conditions, similar to that suggested for the P^c allele in the Mediterranean region (Bottini et al, 1972).

The frequency of the β^S allele is relatively high (16%) among those Beti that have come from the savannah area with the highest malaria endemicity, while it is low among the Babenga living in the forest (4%), a frequency comparable with that in other samples of the Babenga pygmies (Santachiara-Benerecetti et al, 1980).

The heterozygosity index was calculated for each locus, and from these was estimated the average index value for each population for all loci, using Nei and Roychoudhury's formula (1974a). For this purpose unpublished gene frequencies of additional markers (EsD, Gc, Pi, Tf) were also included in order to improve the reliability of the estimate. The average group heterozygosity appears to be quite high, about 27% among Beti and north Bateke, 23% among south Bateke and about 24% among Babenga (Table 3). The northern Bateke group has a heterozygosity more similar to the Beti than to the southern Bateke; the latter, in the territory where they were first established, maintain a degree

Table 1: Erythrocyte and serum phenotypes in Beti, Bateke and Babenga

Locus	Phenotypes	Beti		Bateke (North)		Bateke (South)		Babenga	
		No. Obs.	No. Exp.	No. Obs.	No. Exp.	No. Obs.	No. Exp.	No. Obs.	No. Exp.
6PGD	A	88	88.13	79	79	56	56	46	46
	AC	7	6.73	-		1	0.99	-	
	C	0	0.13	-		0	0.004	-	
	Total	95		79		57		46	
	χ^2_1	0.14			-	0.004			-
AcP	A	3	3.60	4	2.63	0	0.53	0	1.72
	B	59	59.21	55	53.63	44	44.74	29	30.72
	AB	30	29.21	21	23.73	11	9.75	18	14.55
	RA	1	0.58	-	-	0	0.19	-	
	RB	2	2.37	-	-	2	1.77	-	-
	R	0	0.02	-		0	0.02	-	
	Total	95		80		57		47	
	χ^2_3	0.50		1.06		2.27		2.63	
GLO I	1-1	10	8.55	7	7.81	8	6.01	3	4.17
	2-1	37	39.90	36	34.37	21	24.99	22	19.66
	2-2	48	46.55	37	37.81	28	26.01	22	23.17
	Total	95		80		57		47	
	χ^2_1	0.51		0.18		1.44		0.67	
Hbβ	A	64	66.52	66	66.61	46	46.53	43	43.08
	AS	31	25.95	14	12.77	11	9.94	4	3.83
	S	0	2.53	0	0.61	0	0.53	0	0.08
	Total	95		80		57		47	
	χ^2_1	3.61		0.73		0.65		0.09	
PGM_1	1-1	54	57.65	51	49.43	45	45.41	29	27.39
	2-1	40	32.71	23	26.12	9	8.22	13	16.22
	2-2	1	4.64	5	3.45	-	0.37	4	2.39
	Total	95		79		54		46	
	χ^2_1		4.67		1.17		0.06		1.81
Hp	1-1	38	36.04	24	22.47	22	20.24	12	11.87
	2-1	32	35.98	26	29.10	15	18.52	15	15.24
	2-2	11	8.98	11	9.42	6	4.24	5	4.89
	Total	81		61		43		32	
	χ^2_1	1.00		0.70		1.55		0.01	
Km	1-	11	11.25	11	10.48	7	10.87	3	4.79
	1-3	43	42.54	36	36.89	34	26.26	21	17.40
	-3	40	40.20	33	32.46	12	15.86	14	15.81
	Total	94		80		53		38	
	χ^2_1	0.01		0.06		4.60		1.62	

AK, SOD A, and Cp loci were invariant among the four populations.
Two 6-PGD variants were excluded from the figure.

Table 2: Erythrocyte and serum polymorphisms examined: gene frequencies

Locus	Alleles	Beti		Bateke (North)		Bateke (South)		Babenga	
		Freq.	S.E.	Freq.	S.E.	Freq.	S.E.	Freq.	S.E.
6PGD	PGD^A	.963		1.000		.991		1.000	
	PGD^C	.037	.014	-		.009	.009	-	
AcP	P^a	.195	.029	.181	.030	.097	.028	.191	.041
	P^b	.789	.030	.819		.886	.030	.809	
	P^r	.016	.009	-		.017	.012	-	
PGM_1^2	PGM^2	.221	.030	.209	.032	.083	.027	.228	.043
GLO I	GLO^1	.300	.033	.312	.037	.325	.044	.298	.047
Hbβ	β^S	.163	.027	.087	.022	.096	.028	.043	.021
*Hp	Hp^2	.333	.037	.393	.044	.314	.050	.391	.061
Km	Km 1	.346	.035	.362	.038	.453	.048	.355	.055

* Ahaptoglobinemic individuals (Hp O) were excluded from gene counting.

Table 3: Average heterozygosity

Locus	Beti		Bateke (North)		Bateke (South)		Babenga	
	h	V_h	h	V_h	h	V_h	h	V_h
6PGD	.0709	.0030	-	-	.0174	.0036	-	-
AcP	.3375	.0008	.2967	.0003	.2054	$3x10^{-5}$	.3097	.0007
EsD	.1453	.0010	.1597	.0007	.0344	.0030	.0618	.0026
GLO	.4200	.0031	.4297	.0034	.4385	.0054	.4183	.0038
PGM_1	.3820	.0019	.3515	.0011	.2507	.0002	.4246	.0041
Hbβ	.2731	$8x10^{-5}$	.1597	.0007	.1744	.0001	.0816	.0020
Hp	.4444	.0040	.4773	.0054	.4308	.0050	.4761	.0064
Gc	.2312	$2x10^{-5}$	.3142	.0005	.2700	.0004	.2933	.0005
Pi	.1746	.0005	.1737	.0005	.1250	.0007	.0282	.0038
Tf	.1109	.0018	.1744	.0005	.1384	.0005	.1435	.0006
Km	.4526	.0033	.4632	.0036	.4956	.0064	.4579	.0042
mean & SE	.2766 ± .0016		.2727 ± .0013		.2352 ± .0017		.2447 ± .0022	

of heterogeneity similar to that of the native Babenga. It seems therefore that the migration of the Bateke, by dividing them into two groups, has affected their genetic structure.

REFERENCES

Alexandre, P. (1981). Les Africains. Paris: Lidis.

Atlas de la Republique Populaire du Congo (1972). Paris: Jeune Afrique.

Bark, J.E., Harris, M.J. & Firth, M. (1976). Typing of the common phosphoglucomutase variants using isoelectric-focusing. A new interpretation of the phosphoglucomutase system. Journal of the Forensic Sciences Society, **16**, 115-120.

Bender, K., Frank, R., Hitzeroth, H.W. (1977). Glyoxalase I polymorphism in South African Bantu-speaking negroids. Human Genetics, **38**, 223-226.

Bottini, E., Lucarelli, P., Bastianon, V. & Gloria, F. (1972). Erythrocyte acid phosphatase polymorphism and hemolysis. Journal of Medical Genetics, **9**, 434-435.

Carter, N.D., Fildes, R.A., Fitch, L.I. & Parr, C.W. (1968). Genetically determined electrophoretic variations of human phosphogluconate dehydrogenase. Acta Genetica, **18**, 109-122.

Fildes, R.A. & Parr, C.W. (1963). Human red cell phosphogluconate dehydrogenase. Nature, **200**, 890.

Fildes, R.A. & Harris, H. (1966). Genetically determined variation of adenylate kinase in man. Nature, **209**, 261-263.

Golias, T.L. (1971). Helena Laboratories Electrophoresis Manual. Beaumont, Texas: Helena Laboratories.

Hiernaux, J. (1976). Blood polymorphism frequencies in the Sara Majingay of Chad. Annals of Human Biology, **3**(2), 127-140.

Hitzeroth, H.W., Bender, K. & Frank, R. (1981). South African negroes: isoenzyme polymorphisms (GPT, PGM_1, PGM_2, AcP, AK and ADA) and tentative genetic distances. Anthropologischer Anzeiger, **1**, 20-35.

Hopkinson, D.A., Spencer, N. & Harris, H. (1963). Red cell acid phosphatase variants: a new human polymorphism. Nature, **199**, 969.

Jenkins, T. & Corfield, V. (1972). The red cell acid phosphatase polymorphism in southern Africa: population data and studies on the R, RA and RB phenotypes. Annals of Human Genetics, **35**, 379.

Jenkins, T. & Nurse, G.T. (1974). The red cell 6-phosphogluconate dehydrogenase polymorphism in certain southern African populations, with the first report of a new phenotype. Annals of Human Genetics, **38**, 19.

Ki Zerbo, J. (1972). Histoire de l'Afrique noire. D'hier a demain. Paris: Hatier.

Kompf, J., Bissbort, S., Gussmann, S. & Ritter, H. (1975). Polymorphism of red cell glyoxalase I. A new genetic marker in man. Humangenetik, **27**, 141-143.

Le Gall, J.Y., Le Gall, M., Godin, Y. & Serre, J.L. (1982). A study of genetic markers of the blood in four central African population groups. Human Heredity. **32**. 418-427.

Mair, L. (1977). African Kingdoms. Oxford: Clarendon Press.

Martin, G., McAlister, R., Pelter, W. & Benditt, E.P. (1961). Heterogeneity of ceruloplasmin. Proceedings of the Second International Congress of Human Genetics, **2**, 752.

Meera Khan, P. (1971). Enzyme electrophoresis on cellulose acetate gel: zymogram patterns in man-mouse and man-Chinese hamster somatic cell hybrids. Archives of Biochemistry and Biophysics, **145**, 470-483.

Nei, M. & Roychoudhury, A.K. (1974a). Sampling variances of heterozygosity and genetic distances. Genetics, **76**, 379-390.

Nei, M. & Roychoudhury, A.K. (1974b). Genetic variation within and between the major races of man, Caucasoids, Negroids and Mongoloids. American Journal of Human Genetics, **26**, 421-443.

Nurse, G.T. & Jenkins, T. (1982). Serogenetic studies on the Basters of Rehoboth, south-west Africa/Namibia. Annals of Human Biology, **9**, 157-166.

Nurse, G.T., Jenkins, T., Santos David, J.H. & Steinberg, A.G. (1979). The Njinga of Angola: a serogenetic study. Annals of Human Biology, **6**, 337-348.

Oliver, R. & Fage, J.D. (1972). A short history of Africa. Middlesex: Penguin Books.

Ross, J.A. (ed.) (1982). International Encyclopedia of Population, vol. 2. New York.

Santachiara-Benerecetti, A.S., Beretta, M., Negri, M., Ranzani, G., Antonini, G., Barberlo, C., Modiano, G., Cavalli-Sforza, L.L. (1980). Population genetics of red cell enzymes in pygmies: a conclusive account. American Journal of Human Genetics, **32**, 934-954.

Smithies, O. (1955). Zone electrophoresis in starch gels: group variations in the serum proteins of normal adults. Biochemical Journal, **61**, 629-641.

Snedecor, G.W. & Cochran, W.G. (1972). Statistical methods. The Iowa State University Press.

Spedini, G., Capucci, E., Fuciarelli, M. & Rickards, O. (1980). The AcP polymorphism frequencies in the Mbugu and Sango of Central Africa (correlation between the p^r allele frequencies and some climatic factors in Africa). Annals of Human Biology, **7**, 125-128.

Spencer, N., Hopkinson, D.A. & Harris, H. (1964). Phosphoglucomutase polymorphism in man. Nature, **204**, 742.

Vallois, H.V. & Marquer, P. (1980). Notes ethno-demographiques sur les pygmees de la Republique Populaire du Congo et de ses zones frontieres avec les Republiques Centrafricaine et Gabonaise. Bulletins et Memoires de la Societe d'Anthropologie de Paris, **7**, 109-123.

Vergnes, H., Sevin, A., Sevin, J. & Jaeger, G. (1979). Population genetic studies of the Aka Pygmies (Central Africa). Human Genetics, **48**, 343-355.

POPULATION STRUCTURE STUDIES AND GENETIC VARIABILITY IN HUMANS

L. B. JORDE

Department of Human Genetics, University of Utah School of Medicine, Salt Lake City, Utah, U.S.A.

INTRODUCTION

This short review of human genetic structure deals with genetic distances, display techniques, F-statistics, new types of genetic data, and genetic data other than gene frequencies (migration, isonymy, anthropometry, and pedigrees). This treatment will not be comprehensive, and there are already several other reviews, some of which are more detailed (e.g., Cannings & Cavalli-Sforza, 1973; Fix, 1979; Goodman, 1974; Gower, 1972; Harpending, 1974; Howells, 1973; Jorde, 1980, 1985; Lalouel, 1980; Leslie, 1985; Relethford & Lees, 1982; Roberts, 1975; Smith, 1977; Swedlund, 1980). In addition, applications of many of these methods are contained in the volume by Crawford and Mielke (1982). Since the mathematical equations underlying these methods are readily available in other reviews and in the original papers, most are not repeated here.

GENETIC DISTANCE

Genetic distance measures can be grouped into five broad categories: chi-squared, angular transformation, and gene substitution distances, information measures and non-parametric measures. Each of these will be discussed briefly.

The *chi-squared distances* include those of Sanghvi (1953), Balakrishnan and Sanghvi (1968), Morton et al (1971), Harpending and Jenkins (1973), Reynolds et al (1983), Steinberg et al (1967), and Kurczynski (1970). All of these involve the calculation of a squared difference between gene frequencies in two populations and the standardisation of this difference. These approaches are most satisfactory when the differences between gene frequencies in subpopulations are not too large (Jorde, 1985).

The *angular transformation distances* (Edwards, 1971; Edwards & Cavalli-Sforza, 1972) use an arcsine transformation of gene frequencies in order to make the variances of the frequencies independent of the frequency values. These distances tend to become inaccurate at extreme gene frequencies ($p < .05$ and $p > .95$) (Nei, 1973a), but this is true to some extent of all genetic distance measures (Felsenstein, 1985). Since the random drift model upon which these measures are based assumes equal size of populations, they function best when sample sizes are relatively constant.

Nei (1972, 1973a) and Latter (1973a,b) proposed measures which are designed to measure actual codon differences between populations. These *gene substitution measures* are based on a model which incorporates both mutations and random drift and assumes equilibrium between the two processes. Most other genetic distance measures stem from pure drift models.

These distance measures work well for comparisons at and above the species level, as they tend to be approximately linear with time (Nei, 1975). If a minimal amount of gene flow has occurred between populations, these approaches can be used to estimate divergence times. However, human populations usually undergo far too much gene flow to permit reliable estimates of divergence dates to be made. Unlike other distance measures, the sampling properties of Nei's measures have been studied quite thoroughly (Mueller, 1979; Nei, 1978; Nei & Roychoudhury, 1974). This allows one to calculate standard errors and confidence limits for the distance estimates.

The *information measures* estimate gene diversity within and between populations using information-theoretic techniques. Although these approaches do not provide specific pairwise distances, they do give estimates of the overall extent of genetic distance among populations. The best-known such approach is the Shannon-Wiener index, which was applied by Lewontin (1972) in a study of genetic variation among major human races. He demonstrated that most genetic variation (about 85%) exists within, rather than between, the major races. Similar approaches were put forward by Nei and colleagues (Chakraborty, 1980; Nei, 1973b, Latter, 1980; Rao, 1982). Rao (1982) reviewed all of these measures and found that they give highly concordant results at most gene frequencies. However, the Shannon-Wiener measure does tend to overestimate diversity at low gene frequencies.

Karlin et al (1979, 1982) proposed a set of *non-parametric methods* to measure genetic distance between populations. This approach basically involves comparison of the gene frequencies of each subpopulation to those of a "standard" population (e.g., the average of all subpopulations or a founding population). A nonparametric sign test is used to determine whether one subpopulation is closer to the standard than another. The subpopulations are then ordered with respect to their relative distances from the standard. Statistical significance levels can be assigned to these comparisons.

Although a large number of distance measures exist, they tend to yield very similar results when compared (Balakrishnan, 1974; Felsenstein, 1985; Goodman, 1974; Hedrick, 1975; Jorde, 1980; Krzanowski, 1971; Sanghvi & Balakrishnan, 1972). The nonparametric methods, which are relatively new, have been subjected to less comparative scrutiny. One analysis which compared them with other methods did find a high degree of congruence (Carmelli & Jorde, 1982). The only case in which high correlations were not found involved comparisons at and above the species level (Chakraborty & Tateno, 1976), apparently due to the fact that some measures are more linearly related with time than others.

DISPLAY TECHNIQUES

When genetic distances are calculated for more than a few populations, the size of the resulting matrices can make interpretation quite difficult. Several techniques have been devised to reduce these matrices into a less cumbersome two-dimensional display. Perhaps the most popular such display approach for human population studies is the *genetic map*, which consists of a maplike display of distances between populations (usually in two or three dimensions, although a higher number of dimensions can be used). Most of these maps are generated by principal coordinates analysis (Jorde, 1980; Lalouel, 1980). New interpolation techniques and colour graphics were used by Menozzi et al (1978) and Piazza et al (1981).

Genetic maps can also be produced by multidimensional scaling methods. These approaches, which have the advantage of statistical robustness, were reviewed by Lalouel (1980).

Evolutionary trees are a second means for displaying genetic distance in a reduced number of dimensions. Some of the earliest such trees were those of Edwards and Cavalli-Sforza (1964) who used a "minimum evolution" approach to estimate topologies and branching points. This criterion is equivalent to the maximum parsimony principle: the "best" tree is the one which requires the fewest number of gene substitutions between branching points. Another often-used criterion used in generating trees is the compatibility principle: the "best" tree is the one which agrees reasonably with the largest number of variables (e.g.

gene frequencies). The compatibility and parsimony principles were both reviewed thoroughly by Felsenstein (1982).

Maximum likelihood methods have also been used in tree estimation (Thompson, 1973; Felenstein, 1973, 1981a,b). Here the best tree is the one in which the statistical likelihood of a given data set is maximised, given a particular phylogenetic model. Although maximum likelihood estimation provides a firm statistical basis for deriving trees, it requires a comparatively large amount of computer time and involves statistical assumptions that are sometimes untenable. Also, local maxima are sometimes encountered in the likelihood surface; thus, there is no guarantee that the truly "best" tree will be found.

Templeton (1983a,b) proposed a nonparametric approach to generating evolutionary trees. His method, which was designed for use with DNA restriction site or sequence data, has the advantages of distribution-free robustness and relatively rapid computation.

Unlike genetic maps, trees offer the potential to estimate divergence times between populations. However, as noted above, gene flow between populations complicates such estimates considerably. Lathrop (1982) presented a method that provides maximum likelihood estimates of both the phylogeny and the gene flow rates between populations. The method assumes that the duration of admixture is short compared to the period of isolation between populations. Although this assumption precludes the use of this method in many human populations, it should be applicable in at least some cases.

A third approach to the display of genetic distances consists of computing *autocorrelations* of gene frequencies (Sokal & Friedlaender, 1982; Sokal & Menozzi, 1982; Sokal & Wartenberg, 1983). An autocorrelation is the correlation of a series of gene frequencies, in this case arranged along a spatial coordinate system, with itself. Autocorrelations at regular geographic distances are plotted against geographic distance in a correlogram. The patterns of autocorrelations can reflect the relative influence of drift, gene flow, and natural selection. The method was applied to genetic data from Bougainville (Sokal & Friedlaender, 1982) and to pan-European gene frequencies to test hypotheses about the spread of farming during the Neolithic period (Sokal & Menozzi, 1982).

ANALYSIS OF GENETIC DIFFERENTIATION: F-STATISTICS

F statistics were originated by Sewall Wright (1943, 1951, 1965, 1973). They involve the estimation of inbreeding within subpopulations (F_{IS}), inbreeding in the total population (F_{IT}), and inbreeding due to the division of the population into subdivisions (F_{ST}). When using gene frequencies, the latter quantity is estimated by

$$F_{ST} = \frac{Var(p)}{\bar{p}(1-\bar{p})}$$

where p represents gene frequencies in subpopulations and $\bar{p}$ is the mean gene frequency across subpopulations. F_{ST} has been employed extensively to measure genetic differentiation in human populations, since it is essentially a standardised measure of gene frequency variance. A large table of comparative F_{ST} values in various populations has been compiled (Jorde, 1980).

In addition to the method given by equation 1, several other approaches have been proposed to estimate F_{ST} (Nei, 1973b, 1977; Cockerham, 1969, 1973; Rothman et al, 1974). Spielman et al (1977) compared these techniques and demonstrated that all give highly similar results. F_{ST} is also closely related to the heterogeneity chi-squared statistic used by Workman and Niswander (1970).

When F_{ST} is based on gene frequencies from a sample of a population, it is biased upward. Several correction factors have been published to correct this bias (Harpending & Jenkins, 1974; Rogers, 1982; Workman et al, 1973).

The sampling variance of F_{ST} was studied by several researchers (Nei & Chakravarti, 1977; Weir & Cockerham, 1984). These estimates are very useful because they allow the investigator to determine whether F_{ST} values in different populations are statistically different. Jorde (1980) noted several problems in making such comparisons: F_{ST} tends to covary with sample sizes and number and type of subdivisions. Fortunately, the methods proposed by Weir and Cockerham (1984) overcome many of these difficulties.

COMPARISONS OF STATISTICS USING DIFFERENT CLASSES OF LOCI

Most genetic analyses are based on blood group and electrophoretic loci; genetic distances and F-statistics are usually estimated by summing across all loci. A few recent studies have systematically compared the results based on different classes of loci (e.g., blood group vs. electrophoretic systems), and the findings have been disquieting. McLellan et al (1984) studied genetic distances between the Utah Mormons and European populations, using the HLA A and B loci, red cell antigens, and electrophoretic systems. They found that the HLA and electrophoretic loci gave quite similar genetic maps, both corresponding well with the geographic locations of populations (the Mormons plotted close to their northern European ancestors). The red cell antigens, however, were less highly correlated with the other systems and with geographic locations. Similar results were obtained in an analysis of worldwide gene frequencies (Ryman et al, 1983), and Nei (1985) compared results based on blood groups, electrophoretic systems, and mitochondrial DNA sequences in a study of variation among major races. The electrophoretic and DNA sequence data were highly correlated with one another, but they showed lower correlations with the blood group data.

The apparent lack of agreement between blood groups and other loci could be due to several factors: (1) Blood groups may be more susceptible to the effects of natural selection than other systems. However, the HLA system is also almost certainly subjected to selection (Bodmer, 1975). (2) Electrophoretic alleles, which reflect differences in the sizes and shapes of proteins, may be a better indicator of actual DNA sequences than the red cell antigens, which are further removed from the DNA sequences. Again, this would not explain why the HLA loci, which are also cell surface antigens, correlate highly with the electrophoretic loci. (3) Unlike the HLA and electrophoretic systems, which both involve only codominant alleles, many of the red cell antigens have complex dominance relationships among alleles. Thus, gene frequency estimation is more subject to error in these loci, especially in small populations where Hardy-Weinberg equilibrium is less likely to obtain. This problem may be especially severe in tropical populations, since they are usually small in number.

Since blood group loci form the basis for many important studies of human population structure, these findings are a cause for concern. They underscore the need to study as many loci as possible, as advocated so often by Nei and colleagues (Nei, 1978; Nei & Roychoudhury, 1974). Also, the jackknifing procedure described by Rao and Boudreau (1984), in which distance analyses are repeated with a different locus omitted in each analysis, can help to pinpoint which loci might be responsible for distortion. Finally, some of the new data provided by molecular genetic technology, to be described next, may offer a general solution to the problem of limited, unreliable data.

NEW TYPES OF GENETIC DATA

Several new approaches promise to provide large amounts of new genetic data suitable for population structure analysis. Since many of these are discussed in greater detail elsewhere in this volume, they will be treated only briefly here. Potential applications of these new data in human population genetics have been reviewed at length by Jorde (1985).

Many new developments in *electrophoresis* have enabled researchers to detect new alleles. Denaturation by heat or urea, alteration of buffer pH, isoelectric focusing, and modification of gel concentration and composition have all been used to increase the number of observable alleles at a locus (Coyne, 1982). These techniques, however, tend to reveal rare alleles, which are less useful for most population genetic studies.

Two-dimensional electrophoresis has also revealed new types of genetic variation (Garrels, 1983; Goldman & Merrill, 1983; Neel, 1984). Interestingly, most heterozygosity estimates based on this technique are lower than those based on traditional electrophoresis. This may be due to the detection of different classes of proteins by two-dimensional electrophoresis or to differences in the technique itself (Coyne, 1982; Neel, 1984).

A number of new types of DNA data promise to provide important advances in population structure studies. *DNA-DNA hybridisation*, which provides an indirect measure of sequence homology between populations, has been used to study the hominoid phylogeny (Sibley & Ahlquist, 1984). The technique unfortunately does not allow a fine enough resolution of sequence similarity to be used in comparisons between human populations.

DNA sequence data provide a direct indication of genetic differences between populations. Although much work is being done on sequencing nuclear DNA, only a tiny fraction has thus far been sequenced. Mathematical techniques have been proposed to estimate heterozygosity (Tajima, 1983), genetic distance (Kaplan & Risko, 1982; Tajima, 1983), and evolutionary trees (Felsenstein, 1981b, 1983) from DNA sequences. Nei (1985) compared the error rates in estimating trees based on amino acid sequences, electrophoretic loci, restriction site polymorphisms, and DNA (mitochondrial) sequences. The smallest error rate was given by the DNA sequence data.

Restriction site polymorphisms, discussed elsewhere in this volume, provide a useful indirect measure of differences in DNA sequences. Nei and Li

(1979) published early methods for estimating heterozygosity and genetic distances using restriction site data. Nei and Tajima (1981) provided corrections and extensions to this work, and they later formulated a maximum likelihood approach to estimating nucleotide substitution rates (Nei & Tajima, 1983). Other methods for estimating substitution rates (Kaplan & Risko, 1981) and heterozygosity (Ewens et al, 1981) yield similar results. Methods have also been devised to estimate polymorphism levels using restriction site data (Engels, 1981; Ewens et al, 1981; Hudson, 1982).

DNA data offer several important advantages over traditional data for studies of genetic structure. There are now over 200 restriction site loci (see Cooper and Schmidtke (1984) for a partial enumeration), and this number is rapidly increasing. Since virtually any part of the genome can be accessed using these approaches, it should be possible to obtain a truly random sample of the genome for analysis. And finally, DNA sequences with different functions can be studied (e.g., introns, which may be selectively neutral, vs. exons, which code for proteins).

Another useful new source of genetic data is *mitochondrial DNA* (mtDNA). These data are particularly interesting in that mtDNA is inherited only through the maternal line and has a substitution rate that is approximately ten times that of nuclear DNA (see Avise and Lansman, 1983 and Brown, 1983 for reviews). The techniques cited above for nuclear DNA sequences and restriction sites can also be applied to mtDNA sequences and restriction sites. These data have been used in studies of the hominoid phylogeny (Brown et al, 1982; Nei, 1985; Templeton, 1983a) and race-level variation in humans (Brown, 1980; Cann et al, 1982; Denaro et al, 1981; Johnson et al, 1983; Nei, 1985). A potential problem in using mtDNA in human studies is that the standard errors of estimated genetic distances tend to be larger than the distances themselves (Nei, 1985). This is because the mitochondrial genome contains very few genes. Although some of the results obtained for race-level comparisons may thus be suspect, mtDNA will still be useful at least for species-level comparisons such as the hominoid phylogeny.

An interesting counterpart to the mtDNA data is provided by the DNA polymorphisms recently detected on the Y chromosome (Page et al, 1982, 1984). These polymorphisms are, of course, paternally inherited, and, as with mtDNA, they are not subject to recombination. There are currently five such polymorphisms (D. Page, personal communication), but the number should grow.

KINSHIP ESTIMATION USING NON-BIOCHEMICAL DATA

Thus far, this review has dealt with estimates of genetic structure based on biochemical genetic data. Several other types of data, some of which are easier to collect, can also be used to infer genetic structure.

Migration data

Probably the most commonly studied non-biochemical source of genetic data in humans comes from records of migration; some applications are illustrated by Boyce's 1984 volume. Often, civil and parish registers have documented population movements for many years. These records allow the estimation of gene flow, a key process in genetic structure. This can be combined with data on population size, which indicate levels of random drift. Together, the processes of gene flow and drift account for most of human microevolution.

Two relatively simple discrete-distribution migration models which permit analytical solutions are the *island* and *stepping-stone* models. Both entail a number of restrictive assumptions (see Jorde, 1980), so their application in human populations has been rather limited. The island model has been used to estimate F_{ST} in Ireland (Tills, 1977), and the stepping-stone model has also been applied to Irish demographic data (Relethford, 1980).

One continuous-distribution migration model that has been applied in human population studies is Wright's (1943, 1951) "*neighbourhood model*". This model assumes a continuously distributed population and a normal distribution of parent-offspring migration distances. Examples of applications, and criticisms of the assumptions, are given by Harrison and Boyce (1972), Jorde (1980), and Roberts (1975).

A more frequently used continuous migration model is *Malecot's isolation-by-distance model* (Malecot, 1948, 1950, 1959). This model predicts a negative exponential relationship between kinship and geographic distance between populations:

$$\phi(d) = ae^{-bd} \qquad (2)$$

where ϕ is the kinship coefficient for a pair of populations, d is the geographic distance separating the populations, and a and b are estimated parameters. Random local inbreeding is shown by a, while b represents the rate of decay of kinship with increasing distance. Applications of this model to human populations, including a large table of comparative a and b values, were discussed by Jorde (1980). The method has been the subject of some controversy (Felsenstein, 1975, 1979; Lalouel, 1977, 1979).

The *migration matrix* methods are perhaps the most commonly used set of techniques in studies of human gene flow. There are several closely related approaches (Bodmer & Cavalli-Sforza, 1968; Hiorns et al, 1969; Imaizumi et al, 1970; Smith, 1969), all of which use a full migration matrix which specifies the probability that a gene in one subpopulation will be transferred to another by gene flow. Random drift is measured by effective population size. The methods produce a matrix of kinship coefficients, which specify the probabilities that a gene selected from one subpopulation will be identical by descent to a gene selected at the same locus from another. The assumptions, advantages, and disadvantages of each model were reviewed in detail elsewhere (Jorde, 1980).

It is best to use parent-offspring migration data in these applications, although marital migration data, which are easier to obtain, are often used. Although the use of marital migration data involves several assumptions, a recent analysis shows that it should usually produce results similar to those of parent-offspring data (Jorde, 1984).

Migration models assume that migrants are a random sample of the population from which they came. This assumption is often violated. In the Semai Senoi of Malaysia (Fix, 1978) and in South American swidden groups (Neel, 1967; Smouse et al, 1981), kin groups often migrate together. The result is that gene flow does not have as strong a homogenising effect as it does under random gene flow. This "kin-structured migration" appears to be especially common in undeveloped tropical populations, where kin groups are often more important than in technologically "developed" populations (Leslie, 1985).

One of the most interesting recent developments in the use of migration data involves predicting the relationship between covariance matrices based on gene frequencies and those based on gene flow data. Perhaps the best-known such study is that of Harpending and Ward (1982) who used linear algebra to demonstrate a direct and predictable relationship between a genetic covariance matrix and an estimated symmetric Markovian migration matrix. If systematic pressure (immigration from outside the population) is uniform across subpopulations, their theory predicts a linear relationship between heterozygosity in each subpopulation and the subpopulation's distance from the gene-frequency centroid (given by $(p_i - \bar{p})^2/\bar{p}(1-\bar{p})$). When heterozygosity is plotted against distance from the centroid, those subpopulations that deviate from the regression line are inferred to have received more or less than average immigration from outside the population. Such plots have been used in several genetic distance studies (Crawford & Devor, 1980; Crawford et al, 1981; Devor et al, 1984).

Wijsman et al (1984) used a multivariate least-squares regression approach to predict a migration matrix from matrices of genetic covariance in different time periods. Although their method was formulated to deal with covariance matrices based on isonymy, it can also be used for gene-frequency covariance matrices. Rogers (1982) predicted total migration rates and effective sizes using differences in F_{ST} values between parents and offspring, and Morton (1982) used the Malecot isolation-by-distance model given by equation 2 to predict effective population sizes and migration rates. All of these approaches offer useful potentials for testing predictions generated by evolutionary theory.

Isonymy

Because of their ready availability in historical records, isonymy (same-name marriage) data have been used extensively in human genetic structure studies, on which Gottlieb (1983) edited a recent collection of papers. Isonymy data can be used to estimate both inbreeding (Crow and Mange, 1965) and between-subdivision kinship (Morton, 1971; Lasker, 1977). Certainly the most commonly-voiced criticism of isonymy data is the fact that surnames have multiple ancestral origins (polyphyletism). Thus, isonymy tends to overestimate inbreeding as well as kinship between subdivisions (Lasker, 1978).

The method of Wijsman et al (1984), cited above, was devised to predict a migration matrix using isonymy covariance matrices in different time periods. The other methods cited in that section could be used in the same way.

Since surnames are inherited paternally in many societies, interesting comparisons might be made between population structure estimated using isonymy and structure estimated using Y chromosome polymorphisms. Similarly, a comparison with mtDNA polymorphisms might be useful in populations in which surnames are inherited through the maternal line.

Pedigrees

Although it is difficult to assemble pedigree information of any depth, this has been accomplished for a few populations: the Ramah Navajo (Spuhler &

Kluckhohn, 1953), Tristan da Cunha (Roberts, 1967), the Utah Mormons (Skolnick, 1980), and Tokelau (Ward, 1980) are examples. Pedigree data provide direct measures of inbreeding, and, if available on multiple subdivisions, they can be used to estimate kinship coefficients between subdivisions. Because they usually extend only for a few generations, pedigree data generally underestimate inbreeding and kinship coefficients.

Anthropometric data

Anthropometric data, including dermatoglyphics and body and dental measurements, have been obtained on many populations, and a recent review of many of these studies is available (Relethford & Lees, 1982). Generally, the correlations between anthropometrics and biochemical genetic data have not been high, particularly at the level of major races (Nei & Roychoudhury, 1972, 1982; Piazza et al, 1975). This may be due to the apparent fact that morphological traits are usually more responsive to the effects of natural selection than are most biochemical gene loci.

It has been argued that anthropometric data may provide a more meaningful assessment of population structure than biochemical genetic data (Froehlich & Giles, 1981; Relethford & Lees, 1982; Rothhammer et al, 1977; Workman & Niswander, 1970). This assertion appears to derive from Birdsell's (1950) argument that genetic drift should have less impact on polygenic traits than on monogenic traits, since the effects of drift at multiple loci should cancel one another. The polygenic traits would thus be less subject to stochastic "noise" and would give a more reliable indication of true historical relationships between populations. Although this argument has intuitive appeal, it is based on inappropriate reasoning (Rogers & Jorde, in preparation). The quantity of interest for this question is the between-groups genetic variance. Wright (1951) demonstrated that the expected value of the between-groups variance is identical in monogenic and polygenic traits. Rogers and Harpending (1983) extended this analysis to show that the variance of the between-groups variance (and thus the precision of the estimate) is also equivalent for monogenic and polygenic traits. Thus, monogenic traits are at least as useful as polygenic traits for studies of population structure. In addition, because most polygenic traits have heritability less than one, and because polygenic traits are more highly intercorrelated with one another, there is probably less unique genetic information in a set of polygenic traits than in a set of monogenic traits.

CONCLUSIONS

As this review has shown, there are many useful methods that can be used to study the genetic structure of tropical populations. Useful data also exist, although blood groups, a mainstay of population genetic analysis, have been called into question. Many of the problems inherent in the current repertoire of biochemical loci will be overcome by using DNA polymorphisms.

Anthropometric data, although perhaps not quite as useful as monogenic data, can easily be collected for tropical populations and thus also offer potential. Since tropical populations tend to have fewer written records than many other populations, the opportunities for collecting migration, isonymy, and pedigree data tend to be somewhat fewer. However, at least one generation of parent-offspring migration data can be collected orally, and it has been possible to assemble impressively large pedigrees for some tropical populations (Ward, 1980). These ancillary types of data can provide very interesting and informative comparisons with biochemical data.

Nearly all of the techniques and types of data discussed in this review can be useful in studies of tropical populations. Because of their many unique historical and adaptive features, tropical populations can contribute considerably to our understanding of human evolution.

ACKNOWLEDGMENTS

I am grateful for comments from Drs. Dorit Carmelli, Mary Dadone, John Endler, Henry Harpending, Mark Leppert, Tracy McLellan, Kenneth Morgan, David Page, Alan Rogers, Richard Sage, and Ray White. Any errors are my own. Financial support for this work was provided by NIH grant HD-16109 and NSF grant BNS-8319448.

REFERENCES

Avise, J. C. & Lansman, R. A. (1983). Polymorphism of mitochondrial DNA in populations of higher animals. In: M. Nei (ed.), Evolution of Genes and Proteins, pp. 147-164. Sunderland, Mass.: Sinauer.

Balakrishnan, V. (1974). Comparison of some commonly-used genetic distance measures. In: H.M. Bhatia, P.K. Sukumaran & J.V. Undevia (eds.), Human Population Genetics in India, pp. 173-86. New York: Orient Longman.

Balakrishnan, V. & Sanghvi, L. D. (1968). Distance between populations on the basis of attribute data. Biometrics, **24**, 859-865.

Birdsell, J. B. (1950). Some implications of the genetical concept of race in terms of spatial analysis. Cold Spring Harbor Symposium on Quantitative Biology, **15**, 259-314.

Bodmer, W. F. (1975). Evolution of HL-A and other major histocompatibility systems. Genetics, **79**, 293-304.

Bodmer, W. F. & Cavalli-Sforza, L. L. (1968). A migration matrix model for the study of random genetic drift. Genetics, **59**, 565-92.

Boyce, A. J. (ed.) (1984). Migration and Mobility: Biosocial Aspects of Human Movement. London: Taylor & Francis.

Brown, W. M. (1980). Polymorphism in mitochonridal DNA of humans as revealed by restriction endonuclease analysis. Proceedings of the National Academy of Sciences, U.S.A., **77**, 3605-3609.

Brown, W. M. (1983). Evolution of animal mitochondrial DNA. In: M. Nei & R. K. Koehn (eds.), Evolution of Genes and Proteins, pp. 62-88. Sunderland, Mass.: Sinauer.

Brown, W. M., Prager, E. M., Wang, A. & Wilson, A. C. (1982). Mitochondrial DNA sequences of primates: tempo and mode of evolution. Journal of Molecular Evolution, **18**, 225-239.

Cann, R. L., Brown, W. M. & Wilson, A. C. (1982). Evolution of human mitochondrial DNA: A preliminary report. In: B. Bonné-Tamir, T. Cohen & R.M. Goodman (eds.), Human Genetics, Part A: The Unfolding Genome, pp. 157-165. New York: Alan R. Liss.

Cannings, C. & Cavalli-Sforza, L. L. (1973). Human population structure. Advances in Human Genetics, **4**, 105-171.

Carmelli, D. & Jorde, L. B. (1982). A nonparametric distance analyais of biochemical genetic data from the Åland Islands, Finland. American Journal of Physical Anthropology, **57**, 331-340.

Chakraborty, R. (1980). Gene diversity analysis in nested subdivided populations. Genetics, **96**, 721-726.

Chakraborty, R. & Tateno, Y. (1976). Correlations between some measures of genetic distance. Evolution, **30**, 851-853.

Cockerham, C. C. (1969). Variance of gene frequencies. Evolution, **23**, 72-84.

Cockerham, C. C. (1973). Analysis of gene frequencies. Genetics, **74**, 679-700.

Cooper, D. N. & Schmidtke, J. (1984). DNA restriction fragment length polymorphisms and heterozygosity in the human genome. Human Genetics, **66**, 1-16.

Coyne, J. A. (1982). Gel electrophoresis and cryptic protein variation. In: M.C. Ratazzi, J.G. Scandalios & G.S. Whitt (eds.), Isoenzymes: Current Topics in Biological and Medical Research, vol. 6, pp. 1-32. New York: Alan R. Liss.

Crawford, M. H. & Devor, E. J. (1980). Population structure and admixture in transplanted Tlaxcaltecan populations. American Journal of Physical Anthropology, **52**, 485-490.

Crawford, M. H. & Mielke, J. H. (eds.) (1982). Current Developments in Anthropological Genetics, Vol. 2, Ecology and Population Structure. New York: Plenum Press.

Crawford, M. H., Meilke, J. H., Devor, E. J., Dykes, D. D. & Polesky, H. F. (1981). Population structure of Alaskan and Siberian indigenous communities. American Journal of Physical Anthropology, **55**, 167-185.

Denaro, M., Blanc, H., Johnson, M. J., Chen, K. H., Wilmsen, E., Cavalli-Sforza, L. L. & Wallace, D. C. (1981). Ethnic variation in Hpa I endonuclease cleavage patterns of human mitochondrial DNA. Proceedings of the National Academy of Sciences, U.S.A., **78**, 5768-5772.

Devor, E. J., Crawford, M. H. & Bach-Enciso, V. (1984). Genetic structure of the Black Caribs and Creoles. In: M. H. Crawford (ed.), Current Developments in Anthropological Genetics, Vol. 3, Black Caribs: A Case Study in Biocultural Adaptation, pp. 365-380. New York: Plenum Press.

Edwards, A. W. F. (1971). Distance between populations on the basis of gene frequencies. Biometrics, **27**, 873-881.

Edwards, A. W. F. & Cavalli-Sforza, L. L. (1964). Reconstruction of evolutionary trees. In: V.E. Heywood & J. McNeill (eds.), Phenetic and Phylogenetic Classification. (Systems Association Publication No. 6). London: The Systematics Association.

Edwards, A. W. F. & Cavalli-Sforza, L. L. (1972). Affinity as revealed by differences in gene frequencies. In: J.S. Weiner & J. Huizinga (eds.), The Assessment of Population Affinities in Man, p. 37-47. Oxford: Clarendon.

Engels, W. R. (1981). Estimating genetic divergence and genetic variability with restriction endonucleases. Proceedings of the National Academy of Sciences, U.S.A., **78**, 6329-6333.

Ewens, W. J., Spielman, R. S. & Harris, A. (1981). Estimation of genetic variation at the DNA level from restriction endonuclease data. Proceedings of the National Academy of Sciences, U.S.A., **78**, 3748-3750.

Felsenstein, J. (1973). Maximum-likelihood estimation of evolutionary trees from continuous characters. American Journal of Human Genetics, **25**, 471-492.

Felsenstein, J. (1975). A pain in the torus: some difficulties with models of isolation by distance. American Nature, **109**, 359-368.

Felsenstein, J. (1979). Isolation by distance: reply to Lalouel and Morton. Annals of Human Genetics, **42**, 523-527.

Felsenstein, J. (1981a). Evolutionary trees from gene frequencies and quantitative characters: finding maximum likelihood estimates. Evolution, **35**, 1229-1242.

Felsenstein, J. (1981b). Evolutionary trees from DNA sequences: a maximum likelihood approach. Journal of Molecular Evolution, **17**, 368-376.

Felsenstein, J. (1982). Numerical methods for inferring evolutionary trees. Quarterly Review of Biology, **57**, 379-404.

Felsenstein, J. (1983). Inferring evolutionary trees from DNA sequences. In: B.S. Weir (ed.), Statistical Analysis of DNA Sequence Data, pp. 133-150. New York: Marcel Dekker.

Felsentstein, J. (1985). Phylogenies from gene frequencies: a statistical problem. Systematic Zoology (in press).

Fix, A. G. (1978). The role of kin-structured migration in genetic microdifferentiation. Annals of Human Genetics, **41**, 329-339.

Fix, A. G. (1979). Anthropological genetics of small populations. Annual Review of Anthropology, **8**, 207-230.

Froehlich, J. W. & Giles, E. (1981). A multivariate approach to fingerprint variation in Papua New Guinea: perspectives on the evolutionary stability of dermatoglyphic markers. American Journal of Physical Anthropology, **54**, 93-106.

Garrels, J. I. (1983). Quantitative two-dimensional gel electrophoresis of proteins. In: R. Wu, L. Grossman & K. Moldave (eds.), Methods in Enzymology, vol. 100, Recombinant DNA, Part B, pp. 411-423. New York: Academic Press.

Goldman, D. & Merrill, C. R. (1983). Human lymphocyte polymorphisms detected by quantitative two-dimensional electrophoresis. American Journal of Human Genetics, **35**, 827-837.

Goodman, M. M. (1974). Genetic distances: measuring dissimilarity among populations. Yearbook of Physical Anthropology, 1973, **17**, 1-38.

Gottlieb, K. (ed.), (1983). Symposium on Surnames as Markers of Inbreeding and Migration. Human Biology, **55**, 209-408.

Gower, J. C. (1972). Measures of taxonomic distance and their analysis. In: J.S. Weiner & J. Huizinga (eds.), The Assessment of Population Affinities in Man, pp. 1-24. Oxford: Clarendon Press.

Harpending, H. C. (1974). Genetic structure of small populations. Annual Review of Anthropology, **3**, 229-243.

Harpending, H. C. & Jenkins, T. (1973). Genetic distances among southern African populations. In: M.H. Crawford & P.L. Workman (eds.), Methods and Theories of Anthropological Genetics, pp. 177-199. Albuquerque: University of New Mexico Press.

Harpending, H. C. & Jenkins, T. (1974). !Kung population structure. In: J.F. Crow & C. Denniston (eds.), Genetic Distance, pp. 137-161. New York: Plenum Press.

Harpending, H. C. & Ward, R. H. (1982). Chemical systematics and human populations. In: M.H. Nitecki (ed.), Biochemical Aspects of Evolutionary Biology, pp. 213-256. Chicago: University of Chicago Press.

Harrison, G. A., Boyce, A. J. (1972). Migration, exchange, and the genetic structure of populations. In: G.A. Harrison & A.J. Boyce (eds.), The Structure of Human Populations, pp. 128-145. Oxford: Clarendon Press.

Hedrick, P. W. (1975). Genetic similarity and distance: comments and comparisons. Evolution, **79**, 362-366.

Hiorns, R. W., Harrison, G. A., Boyce, A. J., Kuchemann, C. F. (1969). A mathematical analysis of the effects of movement on the relatedness between populations. Annals of Human Genetics, **32**, 237-250.

Howells, W. W. (1973). Measures of population distances. In: M.H. Crawford & P.L. Workman (eds.), Methods and Theories of Anthropological Genetics, pp. 159-176. Albuquerque: University of New Mexico Press.

Hudson, R. R. (1982). Estimating genetic variability with restriction endonucleases. Genetics, **100**, 711-719.

Imaizumi, Y., Morton, N. E., Harris, D. E. (1970). Isolation by distance in artificial populations. Genetics, **66**, 569-582.

Johnson, M. J., Wallace, D. C., Ferris, S. D., Rattazzi, M. C., Cavalli-Sforza, L. L. (1983). Radiation of human mitochondria DNA types analyzed by restriction endonuclease cleavage patterns. Journal of Molecular Evolution, **19**, 255-271.

Jorde, L. B. (1980). The genetic structure of subdivided human populations: a review. In: J.H. Mielke & M.H. Crawford (eds.): Current Developments in Anthropological Genetics, vol. 1, pp. 135-208. New York: Plenum Press.

Jorde, L. B. (1984). A comparison of parent-offspring and marital migration data as measures of gene flow. In: A.J. Boyce (ed.), Migration and Mobility: Biosocial Aspects of Human Movement, pp. 83-96. London: Taylor & Francis.

Jorde, L. B. (1985). Human genetic distance studies: present status and future prospects. Annual Review of Anthropology, **14**, 343-373.

Kaplan, N. & Risko, K. (1981). An improved method for estimating sequence divergence of DNA using restriction endonuclease mappings. Journal of Molecular Evolution, **17**, 156-162.

Kaplan, N. & Risko, K. (1982). A method for estimating rates of nucleotide substitution using DNA sequence data. Theoretical Population Biology, **21**, 318-328.

Karlin, S., Carmelli, D. & Bonné-Tamir, B. (1982). Analysis of biochemical genetic data on Jewish populations: III. The application of individual haplotype measurements for intra- and inter-population comparisons. American Journal of Human Genetics, **34**, 50-64.

Karlin, S., Kenett, R. & Bonne-Tamir, B. (1979). Analysis of biochemical data on Jewish populations: II. Results and interpretations of heterogeneity indices and distance measures with respect to standards. American Journal of Human Genetics, **31**, 341-365.

Krzanowski, W. J. (1971). A comparison of some distance measures applicable to multinomial data, using a rotational fit technique. Biometrics, **27**, 1062-1068.

Kurczynski, T. W. (1970). Generalized distance and discrete variables. Biometrics, **26**, 525-534.

Lalouel, J. M. (1977). The conceptual framework of Malécot's model of isolation by distance. Annals of Human Genetics, **40**, 355-360.

Lalouel, J. M. (1979). Comment on Felsenstein's reply to Lalouel and Morton. Annals of Human Genetics, **42**, 529.

Lalouel, J. M. (1980). Distance analysis and multidimensional scaling. In: J.H. Mielke & M.H. Crawford (eds.): Current Developments in Anthropological Genetics, vol. 1: Theory and Methods, pp. 209-250. New York: Plenum Press.

Lasker, G.W. (1977). A coefficient of relationship by isonymy: a method for estimating the genetic relationship between populations. Human Biology, **49**, 489-493.

Lasker, G. W. (1978). Increments through migration to the coefficient of relationship between communities estimated by isonymy. Human Biology, **50**, 235-240.

Lathrop, G. M. (1982). Evolutionary trees and admixture: phylogenetic inferences when some populations are hybridized. Annals of Human Genetics, **46**, 245-255.

Latter, B.D.H. (1973a). Measures of genetic distance between individuals and populations. In: N. E. Morton (ed.): Genetic Structure of Populations, pp. 27-37. Honolulu: University of Hawaii Press.

Latter, B. D. H. (1973b). The estimation of genetic divergence between populations based on gene frequency data. American Journal of Human Genetics, **25**, 247-261.

Latter, B.D.H. (1980). Genetic differences within and between populations of the major human subgroups. American Naturalist, **116**, 220-237.

Leslie, P. W. (1985). Potential mates analysis and the study of human population structure. Yearbook of Physical Anthropology (in press).

Lewontin, R. C. (1972). The apportionment of human diversity. Evolutionary Biology, **6**, 381-398.

Malécot, G. (1948). Les Mathematiques de l'Hérédité. Paris: Masson (translated as The Mathematics of Heredity, San Francisco: Freeman, 1969).

Malécot, G. (1950). Quelques schemas probabilistes sur la variabilité des populations naturelles. Ann.Univ.Lyon Sect.A, **13**, 37-60.

Malécot, G. (1959). Les modèles stochastiques en génétique de population. Publ.Inst.Statist.Univ.Paris, **8**, 173-210.

McLellan, T., Jorde, L. B., Skolnick, M. H. (1984). Genetic distances between the Utah Mormons and related populations. American Journal of Human Genetics, **36**, 836-857.

Menozzi, P., Piazza, A., Cavalli-Sforza, L. L. (1978). Synthetic maps of human gene frequencies in Europeans. Science, **201**, 786-792.

Morton, N. E., Yee, S., Harris, D. E., Lewis, R. (1971). Bioassay of kinship. Theoretical Population Biology, **2**, 507-524.

Morton, N. E. (1982). Estimation of demographic parameters from isolation by distance. Human Heredity, **32**, 37-41.

Mueller, L. D. (1979). Comparison of two methods for making statistical inferences on Nei's measure of genetic distance. Biometrics, **35**, 757-763.

Neel, J. V. (1967). The genetic structure of primitive human populations. Japanese Journal of Human Genetics, **12**, 1-16.

Neel, J. V. (1984). A revised estimate of the amount of genetic variation in human proteins: implications for the distribution of DNA polymorphisms. American Journal of Human Genetics, **36**, 1135-1148.

Nei, M. (1972). Genetic distance between populations. American Naturalist, 106: 283-292.

Nei, M. (1973a). The theory and estimation of genetic distance. In: N. E. Morton (ed.), Genetic Structure of Populations, pp. 45-51. Honolulu: University of Hawaii.

Nei, M. (1973b). Analysis of gene diversity in subdivided populations. Proceedings of the National Academy of Sciences, U.S.A., **70**, 3321-3323.

Nei, M. (1975). Molecular Population Genetics and Evolution. Amsterdam: North Holland Publishing Co.

Nei, M. (1977). F-statistics and analysis of gene diversity in subdivided populations. Annals of Human Genetics, **41**, 225-233.

Nei, M. (1978). Estimation of average heterozygosity and genetic distance from a small number of individuals. Genetics, **89**, 583-590.

Nei, M. (1985). Human evolution at the molecular level. In: Proceedings of the Oji Conference on Population Genetics and Evolution (in press).

Nei, M., Li, W. H. (1979). Mathematical model for studying genetic variation in terms of restriction endonucleases. Proceedings of the National Academy of Sciences, U.S.A., **76**, 5269-5273.

Nei, M., Chakravarti, A. (1977). Drift variances of F_{ST} and G_{ST} statistics obtained from a finite number of isolated populations. Theoretical Population Biology, **11**, 307-325.

Nei, M., Roychoudhury, A. K. (1974). Sampling variances of heterozygosity and genetic distance. Genetics, **76**, 379-390.

Nei, M., Roychoudhury, A. K. (1982). Genetic relationship and evolution of human races. Evolutionary Biology, **14**, 1-59.

Nei, M. & Tajima, F. (1981). DNA polymorphism detectable by restriction endonucleases. Genetics, **97**, 145-163.

Nei, M. & Tajima, F. (1983). Maximum likelihood estimation of the number of nucleotide substitutions from restriction site data. Genetics, **105**, 207-217.

Page, D., de Martinville, B., Barker, D., Wyman, A., White, R., Francke, U. & Botstein, D. (1982). Single-copy sequence hybridizes to polymorphic and homologous loci on human X and Y chromosomes. Proceedings of the National Academy of Sciences, U.S.A., **79**, 5352-5356.

Page, D. C., Harper, M. E., Love, J. & Botstein, D. (1984). Occurrence of a transposition from the X-chromosome long arm to the Y-chromosome short arm during human evolution. Nature, **311**, 119-123.

Piazza, A., Menozzi, P. & Cavalli-Sforza, L. L. (1981). Synthetic gene frequency maps of man and selective effects of climate. Proceedings of the National Academy of Sciences, U.S.A., **78**, 2638-2642.

Piazza, A., Sgaramella-Zonta, L., Gluckman, P. & Cavalli-Sforza, L. L. (1975). The Fifth Histocompatibility Workshop on Gene Frequency Data: a phylogenetic analysis. Tissue Antigens, **5**, 445-463.

Rao, C. R. (1982). Diversity and dissimilarity coefficients: a unified approach. Theoretical Population Biology, **21**, 24-43.

Rao, C. R. & Boudreau, R. (1984). Diversity and cluster analyses of blood group data on some human populations. In: A. Chakravarti (ed.), Human Population Genetics: the Pittsburgh Symposium, pp. 331-362. New York: van Nostrand Reinhold.

Relethford, J. H. (1980). Simulation of the effects of changing population size on the genetic structure of western Ireland. Social Biology, **27**, 53-61.

Relethford, J. H. & Lees, F. C. (1982). The use of quantitative traits in the study of human population structure. Yearbook of Physical Anthropology, **25**, 113-132.

Reynolds, J., Weir, B. S., Cockerham, C. C. (1983). Estimation of the coancestry coefficient: basis for a short-term genetic distance. Genetics, **105**, 767-779.

Roberts, D. F. (1967). The development of inbreeding in an island population. Ciencia e Cultura, **19**, 78-84. (Reprinted in Yearbook of Physical Anthropology (1967), **15**, 150-160.)

Roberts, D. F. (1975). Genetic studies of isolates. In: A.E.H. Emery (ed.), Modern Trends in Human Genetics, vol. 2, pp. 221-269. London: Butterworths.

Rogers, A. R. (1982). Variation of Neutral Characters in Subdivided Populations. Ph.D. dissertation, University of New Mexico, Albuquerque.

Rogers, A. R., Harpending, H. C. (1983). Population structure and quantitative characters. Genetics: **105**, 985-1002.

Rothhammer, F., Chakraborty, R. & Llop, E. (1977). A collection of marker gene and dermatoglyphic diversity at various levels of population differentiation. American Journal of Physical Anthropology, **46**, 51-60.

Rothman, E. D., Sing, C. F., Templeton, A. R. (1974). A model for anlaysis of population structure. Genetics, **78**, 943-960.

Ryman, M., Chakraborty, R. & Nei, M. (1983). Differences in the relative distribution of human gene diversity between electrophoretic and red and white cell antigen loci. Human Heredity, **33**, 93-102.

Sanghvi, L. D. (1953). Comparison of genetical and morphological methods for a study of biological differences. American Journal of Physical Anthropology, **11**, 385-404.

Sanghvi, L. D. & Balakrishnan, V. (1972). Comparison of different measures of genetic distance between human populations. In: J.S. Weiner & J. Huizinga (eds.): The Assessment of Population Affinities in Man, pp. 25-36. Oxford: Clarendon.

Sibley, C. G. & Ahlquist, J. E. (1984). The phylogeny of the hominoid primates, as indicated by DNA-DNA hybridization. Journal of Molecular Evolution, **20**, 2-15.

Skolnick, M. H. (1980). The Utah genealogical data base: a resource for genetic epidemiology. In: J. Cairns, J.L. Lyon & M.H. Skolnick (eds.), Cancer Incidence in Defined Populations; Banbury Report No. 4, pp. 285-297. New York: Cold Spring Harbor Laboratory.

Smith, C.A.B. (1969). Local fluctuations in gene frequencies. Annals of Human Genetics, **32**, 251-260.

Smith, C.A.B. (1977). A note on genetic distance. Annals of Human Genetics, **40**, 463-479.

Smouse, P. E., Vitzthum, V. J., Neel & J. V. (1981). The impact of random and lineal fission on the genetic divergence of small human groups: a case study among the Yanomama. Genetics, **98**, 179-197.

Sokal, R. R. & Friedlaender, J. S. (1982). Spatial autocorrelation analysis of biological variation on Bougainville Island. In: M.H. Crawford & J.H. Mielke (eds.), Current Developments in Anthropological Genetics, vol.2: Ecology and Population Structure, pp. 205-227. New York: Plenum Press.

Sokal, R. R. & Menozzi, P. (1982). Spatial autocorrelations of HLA frequencies in Europe support demic diffusion of early farmers. American Naturalist, **119**, 1-17.

Sokal, R. R. & Wartenberg, D. E. (1983). A test of spatial autocorrelation analysis using an isolation-by-distance model. Genetics, **105**, 219-237.

Spielman, R. S., Neel, J. V. & Li, F.H.F. (1977). Inbreeding estimation from population data: models, procedures, and implications. Genetics, **85**, 355-371.

Spuhler, J. N. & Kluckhohn, C. (1953). Inbreeding coefficients of the Ramah Navaho population. Human Biology, **25**, 295-317.

Steinberg, A. G., Bleibtreu, H. K., Kurczynski, T. W., Martin, A. O. & Kurczynski, E. M. (1967). Genetic studies of an inbred human isolate. In: J.F. Crow & J.V. Neel (eds.), Proceedings of the Third International Congress of Human Genetics, pp. 267-289. Baltimore: Johns Hopkins.

Swedlund, A. C. (1980). Historical demography: applications in anthropological genetics. In: J.H. Mielke & M.H. Crawford (eds.), Current Developments in Anthropological Genetics, vol. 1, pp. 17-42. New York: Plenum Press.

Tajima, F. (1983). Evolutionary relationships of DNA sequences in finite populations. Genetics, **105**, 437-460.

Templeton, A. R. (1983a). Phylogenetic inference from restriction endonuclease cleavage site maps with particular reference to the evolution of humans and the apes. Evolution, **37**, 221-244.

Templeton, A. R. (1983b). Convergent evolution and nonparametric inferences from restriction data and DNA sequences. In: B.S. Weir (ed.), Statistical Analysis of DNA Sequence Data, pp. 151-179. New York: Marcel Dekker.

Thompson, E. A. (1975). Human Evolutionary Trees. Cambridge: Cambridge University Press.

Tills, D. (1977). The use of the F_{ST} statistics of Wright for estimating the effects of genetic drift, selection and migration in populations, with special reference to Ireland. Human Heredity, **27**, 153-159.

Ward, R. H., Raspe, P. D., Ramirez, M. E., Kirk, R. L. & Prior, I.A.M. (1980). Genetic structure and epidemiology: the Tokelau study. In: A.W. Eriksson, H. Forsius, H.R. Nevanlinna, P.L. Workman & R.K. Norio (eds.), Population Structure and Genetic Disorders, pp. 301-325. New York: Academic Press.

Weir, B. S. & Cockerham, C. C. (1984). Estimating F-statistics for the analysis of population structure. Evolution, **38**, 1358-1370.

Wijsman, E., Zei, G., Moroni, A. & Cavalli-Sforza, L.L. (1984). Surnames in Sardinia. II. Computation of migration matrices from surname distributions in different periods. Annals of Human Genetics, **48**, 65-78.

Workman, P. L. & Niswander, J. D. (1970). Population studies on southwestern Indian tribes. II. Local genetic differentiation in the Papago. American Journal of Human Genetics, **22**, 24-49.

Workman, P. L., Harpending, H. C., Lalouel, J. M., Lynch, C., Niswander, J.D. & Singleton, R. (1973). Population studies on southwestern Indian tribes. VI: Papago population structure: a comparison of genetic and migration analyses. In: N.E. Morton (ed.), Genetic Structure of Populations, pp. 166-194. Honolulu: University of Hawaii Press.

Wright, S. (1943). Isolation by distance. Genetics, **28**, 114-138.

Wright, S. (1951). The genetical structure of populations. Annals of Eugenics, **15**, 323-354.

Wright, S. (1965). The interpretation of population structure by F-statistics with special regard to systems of mating. Evolution, **19**, 395-420.

Wright, S. (1973). The origin of the F-statistics for describing the genetic aspects of population structure. In: N.E. Morton (ed.), Genetic Structure of Populations, pp. 3-26. Honolulu: University of Hawaii Press.

INBREEDING AND THE INCIDENCE OF RECESSIVE DISORDERS IN THE POPULATIONS OF KARNATAKA, SOUTH INDIA

A. H. BITTLES[1], A. RADHA RAMA DEVI[2] and N. APPAJI RAO[3]

[1]*Department of Anatomy and Human Biology, King's College, London, U.K.*
[3]*Department of Biochemistry and* [2]*Health Centre,*
Indian Institute of Science, Bangalore, India.

INTRODUCTION

While it is generally accepted that inbreeding in humans can lead to increased incidences of genetically-determined abnormalities, due to the expression of rare, deleterious, recessive genes in the homozygous state, consanguineous marriages are common in many communities, for example, the four southern states of India, Andhra Pradesh, Karnataka, Kerala and Tamil Nadu (Kumar et al, 1967; Rao & Inbaraj, 1977a; Rami Reddy & Chandrasekhar Reddy, 1979; Radha Rama Devi et al, 1981). Although infectious diseases and nutritional disorders are still common in India, their incidences have declined markedly during the last twenty years. Therefore it seems probable that among the populations of South India (Fig. 1), which in the 1981 Census totalled over 164 millions, a transition from an almost exclusively environmental to an increasingly genetic pattern of disease is currently under way, similar to that earlier observed in countries such as Great Britain (Roberts et al, 1970). However the emerging situation in the South Indian states almost certainly will be of greater complexity than in Western countries because of the long inbreeding tradition. Indeed, it has been suggested that the high levels of inbreeding practised by the Dravidian peoples for at least 2,000 years (Centerwall et al, 1969) would have led to the gradual elimination of deleterious lethals and sublethals from the gene pool by segregation (Sanghvi, 1966). This theory has been questioned (Chakraborty & Chakravarti, 1977; Bittles, 1980) but supporting evidence was claimed in large-scale prospective and retrospective studies conducted in Tamil Nadu (Rao & Inbaraj, 1977b, 1979a,b, 1980).

Since little or no information has been available relating to the incidence of genetic disease in the South Indian populations, nor even on the types of disorders that are present, it was agreed that a prospective screening programme

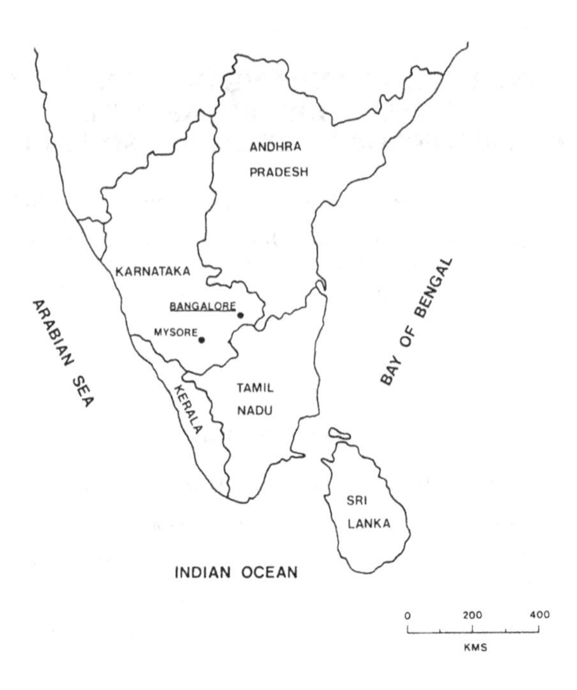

Figure 1: Dravidian South India

of newborn infants should be established. Aminoacid disorders such as phenylketonuria and homocystinuria had been detected in mentally retarded children in Karnataka (Sridhara Rama Rao et al, 1977) and so it was decided that, initially, the screening programme would concentrate on the detection of these conditions.

MATERIALS AND METHODS

The sample collection procedure, in which three drops of blood are taken from the toe or heel on to Whatman 3 MM filter paper, was carried out by trained local staff on babies aged between two and six days born in hospitals in the cities of Bangalore and Mysore, Karnataka. At the same time, details were taken from the mother on her place of residence, age, religion, degree of consanguinity with spouse and the number of her liveborn and living children (Bittles et al, 1982). The aminoacids were eluted overnight into 70% alcohol and analysed in daily batches using a semi-quantitative, thin layer chromatographic technique (Ireland & Read, 1972). Ancillary family data was coded and punched on to cards for storage and subsequent computer analysis. Since the inception of the programme in 1980 almost 60,000 newborns have been tested and the sampling rate is currently 1,200 to 1,400 per month.

RESULTS

Analysis of the first 43,968 pregnancies (Table 1) showed an average consanguinity rate in the current parental generation of 30.2% and coefficient of inbreeding (F) in their infants of 0.02709, which was consistent with the results of the preliminary sample studied in this area (Radha Rama Devi et al, 1982). The levels of inbreeding varied between the major religious groups, being highest among Hindus and lowest in the Christian groups (Table 2). Although the percentage of consanguineous unions was quite similar in Hindus and Muslims, their mean F values differed markedly because of the Hindu preference for uncle-niece marriages (18.2%), as opposed to the first-cousin unions common in the Muslim community (16.2%).

The mean numbers of liveborn and living infants and children in the families of each consanguinity category are shown, with standard errors of the

Table 1: Consanguinity in parents of the newborn population

Total pregnancies studied	43,968		
Non-consanguineous spouses	27,891	63.4%	
Consanguineous spouses			
Beyond second cousin	1,099	2.5%	30.2%
Second cousin	956	2.2%	
First cousin	4,826	11.0%	
Uncle-niece	6,389	14.5%	
Unknown	2,807	6.4%	

Table 2: Religion and consanguinity

Religious group	Incidence in population, %	Consanguineous unions, %	Coefficient of inbreeding, F
Hindu	79.0	34.4	0.03014
Muslim	17.0	25.5	0.01601
Christian	4.0	19.3	0.01435
Other	<0.1	-	-

mean, in Table 3. The values of both variables were marginally higher in each of the inbred groups by comparison with their outbred counterparts. An analysis of variance conducted on the data after arcsin transformation showed no statistically significant relationship between the proportion of survivors and consanguinity.

From the 59,840 newborns tested to date for elevated blood aminoacid levels, a total of forty-two single aminoacidaemias, including hyperphenylalaninaemia, tyrosinaemia, glycinaemia, branched-chain aminoacidaemia and histidinaemia, and forty general aminoacidaemias have been detected. The mean F values in these two groups, 0.03164 and 0.03290 respectively, were little different from that of the total new-born population, 0.02709, slightly higher but not significantly so.

DISCUSSION

Although Bangalore, the state capital, is a modern industrial city, its rapid population expansion during the last thirty years has been largely sustained by the immigration of rural groups from each of the South Indian states. Equally Mysore, the former state capital, has undergone quite rapid industrialisation and growth and so, not surprisingly, the levels of inbreeding observed in the present study are comparable to those reported in other South Indian populations. Almost inevitably the limitation of family sizes, seen in the Hindu and Christian communities, will in time act to restrict the popularity of uncle-niece marriages and hence reduce the overall intensity of inbreeding (Radha Rama Devi et al, 1982). However, contrary to the rapid time-scale earlier envisaged for the abandonment of consanguineous marriages (Dronamraju, 1964), it is now evident that this will be a slow process especially in the predominant rural population.

The failure to demonstrate a significant association between the proportion of survivors and inbreeding was initially surprising but a number of contribu-

Table 3: Consanguinity, liveborn and living children

Marital class	Number studied	No. of liveborns		No. of living children	
		Mean	SEM	Mean	SEM
Non-consanguineous	27,891	2.32	0.0084	2.21	0.0079
Beyond second cousin	1,099	2.49	0.0435	2.38	0.0414
Second cousin	956	2.39	0.0472	2.24	0.0423
First cousin	4,826	2.43	0.0209	2.30	0.0195
Uncle-niece	6,389	2.38	0.0173	2.25	0.0163

tory, interacting factors can be identified:

(i) In the absence of written records relating to the marital patterns of previous generations, their cumulative coefficient of inbreeding values, and thus the total influence of consanguinity on the current gene pool cannot be assessed.

(ii) The present study is cross-sectional and, with respect to survival, retrospective and based on oral report. It tends to be biased towards the ascertainment of living newborns, and gives less evidence on late infant or childhood mortality.

(iii) Studies on completed intra-matrimonial fertility from the neighbouring state of Tamil Nadu have reported first year mortality rates among non-consanguineous individuals to average between 9.3 and 13.2% (Rao & Inbaraj, 1977b, 1979b), suggesting a high, residual "environmental" mortality component against which it is difficult to evaluate the numbers and incidences of genetically-determined disorders.

For these reasons it would be premature to interpret the data on liveborn and living children (Table 3) necessarily in terms of the removal of deleterious genes from the gene pool by long-term inbreeding. This supposition is reinforced by the failure of the aminoacidopathy results either to show a markedly reduced incidence by comparison with Western, outbred levels or a significant increase in the coefficient of inbreeding (F) of the single gene disorder group when compared to the general new-born population, as might be expected if the causative genes were rare in the population (Tchen et al, 1977).

Further evidence that mutant genes are still present in the South Indian gene pools was provided in a separate study on 407 sick infants and children aged twelve days to fourteen years, referred for investigation with a variety of clinical symptoms (Radha Rama Devi et al, in press). In sixty-three of the children, thirty-five distinct genetic disorders were diagnosed: forty-four of the children had single gene defects, twenty-four having a disorder with an autosomal recessive mode of inheritance. As shown in Table 4, there is the clear relationship between inbreeding and the incidence of autosomal recessive disorders that might be expected in these cases.

CONCLUSIONS

Considered specifically in the context of the present volume three main conclusions can be drawn from this study. First, even in human populations with a long history of inbreeding, appreciable levels of genetic variability can be maintained. Second, the investigation of genetic variability via health-based,

Table 4: Consanguinity in sick children

Group studied	Parents consanguineous %	Coefficient of inbreeding, F
All sick children	48.7	0.04135
Single gene defects only	45.2	0.05129
Autosomal recessive defects only	60.9	0.06250
Newborn population	35.2	0.02709

prospective screening programmes already is feasible and of community benefit in developing countries. Finally, the widespread introduction of programmes of this type in other tropical populations whether with similar consanguineous marriage customs or practising outbreeding is both desirable and warranted on biological, social and health grounds.

ACKNOWLEDGMENTS

The generous financial assistance of the British Council, Indian Council for Medical Research, Marie Stopes Research Fund, Overseas Development Administration, and The Royal Society during the course of this work is gratefully acknowledged.

REFERENCES

Bittles, A. H. (1980). Inbreeding in human populations. Journal of Scientific and Industrial Research, **39**, 768-777.

Bittles, A. H., Radha Rama Devi, A., Venkat Rao, S. & Appaji Rao, N., (1982). A new-born screening programme for the detection of aminoacid disorders in South India. Biochemical Reviews, **52**, 20-24.

Centerwall, W. R., Savarinathan, G., Mohan, L. R., Booshanan, V. & Zachariah, M. (1969). Inbreeding patterns in rural South India. Social Biology, **16**, 81-91.

Chakraborty, R. & Chakravarti, A. (1977). On consanguineous marriages and the genetic load. Human Genetics, **36**, 47-54.

Dronamraju, K. R. (1964). Mating systems of the Andhra Pradesh people. Cold Spring Harbor Symposium on Quantitative Biology, **29**, 81-84.

Ireland, J. T. & Read, R. A. (1972). A thin layer chromatographic method for use in neonatal screening to detect excess aminoacidaemia. Annals of Clinical Biochemistry, **9**, 129-132.

Kumar, S., Pai, R. A. & Swaminathan, M. S. (1967). Consanguineous marriages and the genetic load due to lethal genes in Kerala. Annals of Human Genetics, **31**, 141-145.

Radha Rama Devi, A., Appaji Rao, N. & Bittles, A. H. (1981). Consanguinity, fecundity and post-natal mortality in Karnataka, South India. Annals of Human Biology, **8**, 469-472.

Radha Rama Devi, A., Appaji Rao, N. & Bittles, A. H. Inbreeding and the incidence of childhood genetic disorders in Karnataka, South India. Journal of Medical Genetics (in press).

Radha Rama Devi, A., Appaji Rao, N. & Bittles, A. H. (1982). Inbreeding in the States of Karnataka, South India. Human Heredity, **32**, 8-10.

Rami Reddy, V. & Chandrasekhar Reddy, B. K. (1979). Consanguinity effects on fertility and mortality among the Reddis of Chittoor District (A.P.), South India. Indian Journal of Heredity, **11**, 77-88.

Rao, P. S. S. & Inbaraj, S. G. (1977a). Inbreeding in Tamil Nadu, South India. Social Biology, **24**, 281-288.

Rao, P. S. S. & Inbaraj, S. G. (1977b). Inbreeding effects on human reproduction in Tamil Nadu of South India. Annals of Human Genetics, **41**, 87-98.

Rao, P. S. S. & Inbaraj, S. G. (1979a). Inbreeding effects on fertility and sterility in Southern India. Journal of Medical Genetics, **16**, 24-31.

Rao, P. S. S. & Inbaraj, S. G. (1979b). Trends in reproductive wastage in relation to long-term practice of inbreeding. Annals of Human Genetics, **42**, 401-413.

Rao, P. S. S. & Inbaraj, S. G. (1980). Inbreeding effects on fetal growth and development. Journal of Medical Genetics, **17**, 27-33.

Roberts, D. F., Chavez, J. & Court, S. D. M. (1970). The genetic component in child mortality. Archives of Disease in Childhood, **45**, 33-38.

Sanghvi, L. D., (1966). Inbreeding in rural areas of Andhra Pradesh. Indian Journal of Genetics, **26**A, 351-365.

Sridhara Rama Rao, B. S., Subash, M. N. & Reddy Narayanan, H. S. (1977). Metabolic anomalies detected during a systematic biochemical screening of mentally retarded cases. Indian Journal of Medical Research, **65**, 241-247.

Tchen, P., Bois, E., Feingold, J., Feingold, N. & Kaplan, J. (1977). Inbreeding in recessive diseases. Human Genetics, **38**, 163-167.

GENETIC DIVERSITY AT THE ALBUMIN LOCUS

F. W. LOREY[1] and D. G. SMITH[2]

[1]*Department of Anthropology, University of California, Davis, U.S.A.*
[2]*California Primate Research Center, Davis, California, U.S.A.*

INTRODUCTION

Although documentation of genetic diversity in man has been long continuing, the mechanisms maintaining this diversity, in particular selective mechanisms, have proved difficult to identify. The role of balancing selection through heterozygote superiority in fitness was the first to be demonstrated in the resistance to *Plasmodium falciparum* malaria by the sickle cell variant in the haemoglobin polymorphism, and similar mechanisms have been proposed regarding the thalassaemia trait and G6PD deficiency. Other examples of the role of selection in maintaining genetic diversity are still relatively few. Many authorities maintain that the majority of polymorphisms are transient or selectively neutral. Using the assumption of neutralism regarding the albumin molecule, Sarich and Wilson proposed a molecular clock to estimate rates of evolution of, and phylogenetic distances between, mammalian species. However, it seems that sufficient accurate fundamental study of the functional differences between proteins has not yet been undertaken to allow this assumption to be accepted. One reason for this is that, until recently, such study may have been hindered by lack of sophisticated technical equipment. With the advent of more accurate spectrophotometry and, especially, fast protein liquid chromatography (FPLC) which allows for isolation of the protein in question, such study of functional differences in proteins has become possible.

ALBUMIN VARIATION

Albumin is of interest because the very conservative evolution of the molecule, and the rarity of albumin polymorphisms in primates, point to a possible special nature to those albumin polymorphisms which do exist. There is also a strong geographic specificity to the distribution of the polymorphic variants. Table 1 summarises some of the well-known albumin polymorphisms in American Indian populations (Melartin & Blumberg, 1966, 1967; Tanis et al, 1974; Arends et al, 1970; Franklin et al, 1980).

Table 1: Albumin polymorphisms in American Indian populations

Tribe	Variant albumin	Frequency of the variant allele
Naskapi	Naskapi	.130
Navaho	Naskapi	.032
	Mexico	.006
Apache	Naskapi	.015
	Mexico	.034
Pima	Mexico	.049
Papago	Mexico	.013
Hopi	Mexico	.006
Yanomama	Yanomama II	.07-.40

Albumin is an important osmoregulator as well as an important binder of various endogenous substances in the body, as well as drugs. One of the most important catabolites that albumin binds is bilirubin which, in concentrations higher than can be neutralised by the body's normal processes, can infiltrate the tissues to toxic levels and severely damage the central nervous system. Two bilirubin parameters are of relevance. One is the bilirubin-binding capacity (n), the number of binding sites per albumin molecule. Another is the binding association constant (k), the capacity × affinity. This is used to estimate the total biological effectiveness, the binding association constant × the albumin concentration.

These measures of bilirubin binding were analysed on 41 samples of whole Yanomama sera donated by Dr. James Neel (Lorey, 1984). The results indicated that the bilirubin binding capacity was significantly higher in common A/A homozygotes than in the Yan 2 homozygotes, and that the total biological binding effectiveness was also higher in common A/A homozygotes. However, the variance of the Yan homozygotes was significantly lower than that of the common homozygotes, and there was a much narrower distribution of binding capacity and effectiveness. Also, there appeared to be a nonlinear effect of the variant allele in heterozygotes. Variant homozygotes, heterozygotes, and AA homozygotes were found to have mean capacities of 0.89, 1.98 and 2.16 respectively, and affinity constants of 26.5, 11.32, and 9.30 respectively. The sample sizes are small, but these data suggest that other albumin variants may also vary in bilirubin-binding parameters in a similar way to those described here for Yan 2. Indeed similar results have been found with a mixture of variants including Mexico 1 and 2, Makiritare, Naskapi and Yan 2.

In view of the apparently better bilirubin-binding effectiveness of the common albumin type, it is unclear how the variant type can persist in the Yanomama. One possible explanation may lie in the differential displacement of bilirubin by derivatives of coumarin. Coumarins are known to have many deleterious effects, including anticoagulant activity and fertility suppression (Feuer, 1982), and are also known to bind competitively at the same binding site as bilirubin (Brodersen, 1978). Wilding (1977) found that at least one human albumin variant had reduced binding of warfarin (a coumarin derivative). Furthermore, many coumarin-containing plants are known to be present in the diet of some American Indian populations with an albumin variant (Raichelson, 1979), and the tonka bean, used by the Yanomama, contains up to 3% of dry weight coumarin (Pound, 1938).

PRIMATE STUDIES

Further evidence may be sought from the non-human primates, useful for biomedical research because they are biochemically and physiologically so similar to humans (Cornelius and Rosenberg, 1983). One of the most commonly used for this purpose is the macaque (genus Macaca); in recent years these have become readily available at many primate centres across the United States as a result of intensification of domestic breeding efforts (Held, 1980). They are useful for study of the albumin polymorphism, for several populations show a variant albumin form at high frequencies (Table 2). There is also a strong geographic component to the frequency distribution. The usual allele in many populations is MacA, the same as the common Papio allele. The variant MacB replaces it as the common allele in Indian populations but not elsewhere in Asia.

Results of binding studies for whole rhesus plasma in 30 individuals of MacA/MacA, MacA/MacB, MacB/MacB phenotype (Fig. 1) showed mean bilirubin binding capacity (n) of 1.00, 1.06, and 1.11 respectively, and affinity constants (k) of 10.53, 10.95, and 8.20 respectively. Differences between phenotypes were not significant, but were consistent in size and direction with earlier results on whole rhesus plasma in a study by Smith and Ahlfors (1981) which were statistically significant. Interpretation of binding by whole plasma may be obscured by any of several factors leading to marked variation in phenotypic differences. Other non-albumin components of the plasma might affect binding irrespective of phenotype. There may be differences in sex, age, weight, etc., of the monkeys. To eliminate the influence of non-albumin serum factors and animal differences we undertook a study to purify the albumin and to fractionate heterozygous albumin into its two allelic components (Fig. 2) so that binding

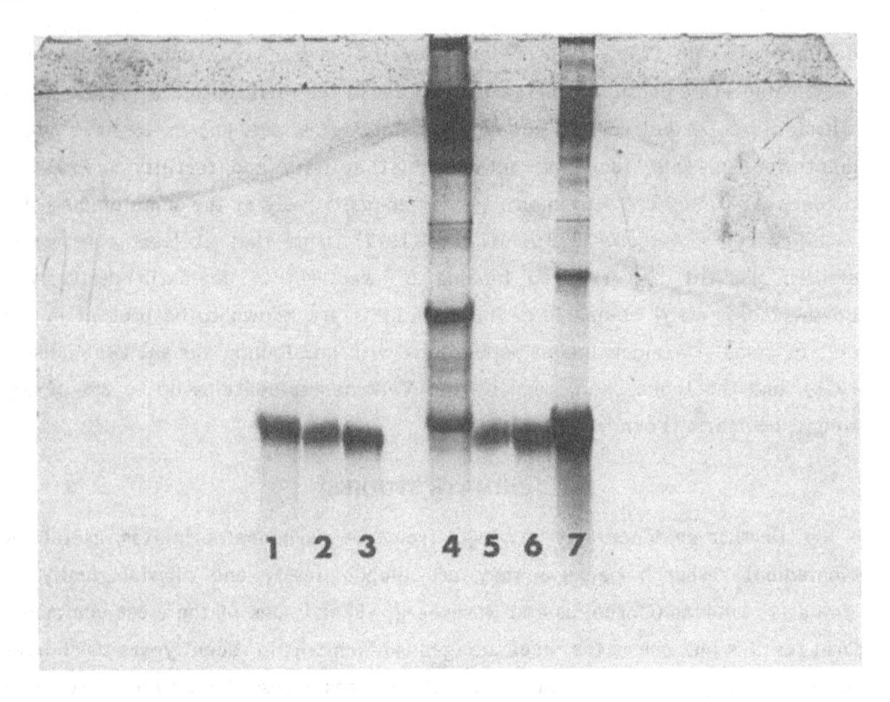

Figure 1: Albumin fractions isolated by chromatography:
1-3: purified macaque albumin of phenotypes A/A, A/B and B/B respectively.
4: Albumin-depleted serum.
5-6: MacA-enriched and MacB-enriched fractions separated by FPLC.
7: Whole serum.

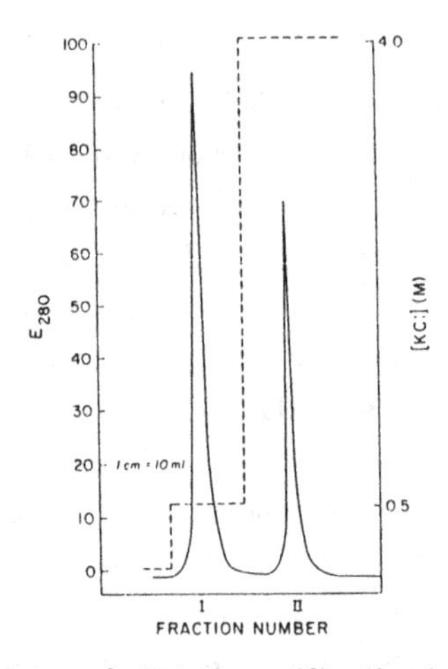

Figure 2: Chromatogram of albumin purification by sepharose:
Elute buffer concentration (KCl) is indicated by the dotted line.

Table 2: Macaque albumin allele frequencies

Species	Location	n	Allele frequency Alb MacA	Alb MacB
M. mulatta	Pakistan	32	0.984	0.016
M. mulatta	Thailand	31	0.952	0.048
M. Mulatta	China	76	0.980	0.020
M. mulatta	India	214	0.400	0.600
M. mulatta	California Primate Center (Indian origin)	456	0.290	0.710
M. radiata	India	19	0	1.0
M. assamensis	India	28	0.161	0.839

analysis could be performed on the two albumin forms within the same animal. This was done by fast protein liquid chromatography, and the subsequent fractionation of the heterozygous albumin by chromatofocusing (Lorey & Smith, 1985).

Blood specimens were obtained from nine rhesus macaques previously identified as heterozygous ($A1^{A}_{mac}/A1^{B}_{mac}$) for the structurally different forms of albumin. The albumin was purified from the specimens, the fractions were analysed, and in each of the nine animals studied the two components were compared. The capacities (n) for eight of the nine comparisons were higher for the variant MacB-enriched fraction. Also in eight of the nine, the values of k and $n \times k$ were higher in the common MacA-enriched portion (Lorey & Smith, 1985). It is difficult to determine which measure, n or k, has more physiological importance. It is tempting to suggest that the MacB variant has increased to such high frequencies because of its higher bilirubin binding capacity (n), but this does not explain why these high frequencies occur only in north Indian populations of rhesus. Again the clue to the geographic specificity of the allele distributions may lie with the coumarin-containing plants. Rhesus macaques in northern India eat several species of plants which may contain coumarin or its derivatives. Two genera of Dipterocarpacea (*Shorea* and *Vatica*), and several genera of the families Apocynacea and Murtaceae, belong to coumarin-containing families or genera of plants, and represent five of the six species of plants most commonly eaten by rhesus in the Siwalik hills of northern India (Lindburg, 1977). Albumin MacB binds bilirubin more effectively than albumin MacA in the presence of coumarin, but less effectively in the absence of coumarin (Ahlfors, unpublished). This may explain why MacB reaches its highest frequencies in areas where coumarins are known to be more common in the diet.

CONCLUSION

More studies of possible differences in function of protein morphs are necessary in the study of genetic diversity. The advent of FPLC technology and more sophisticated spectrophotometry make this possible in many cases. It appears from the several studies using both whole serum and purified albumin, in both humans and rhesus monkeys with albumin variants, that there are indeed significant differences in bilirubin binding between albumin phenotypes. Although more work involving displacers or competitors is required to understand the problem, these initial data suggest that determination of functional differences between proteins may provide useful clues to the selective significance of genetic variants.

REFERENCES

Arends, T., Weitkamp, L., Gallango, M., Neel, J. V., Schultz, J. (1974). Gene frequencies and microdifferentiation among the Makiritare Indians. American Journal of Human Genetics, **22**, 526-532.

Brodersen, R. (1978). Free bilirubin in blood plasma of the newborn. In: L. Stern, W. Oh & B. Fris-Hansen (eds.), Intensive Care in the Newborn, II, pp. 331-345. New York: Masson.

Cornelius, C. E. & Rosenberg, D. P. (1983). American Journal of Medicine, **74**, 169-171.

Feuer, G. (1982). The metabolism and biological action of coumarins, In: R.D.H. Murray (ed.), The Natural Coumarins.

Franklin, S., Wolf, S., Ozdemir, Y., Yuregir, G., Isbir, R., Blumberg, B. (1980). Albumin Naskapi variant in American Indians and Eti Turks. Proceedings of the National Academy of Sciences, **77**, 5480-5482.

Held, J. R. (1980). International Journal of Studies in Animal Problems, **2**, 27-37.

Lindburg, D. (1977). Feeding behavior and diet of rhesus monkeys in a Siwalik forest in North India, In: T.H. Clutton-Brock (ed.), Primate Ecology, pp. 223-249. Academic Press.

Lorey, F. W., Ahlfors, C. E., Smith, D. G. & Neel, J. V. (1984). American Journal of Human Genetics, **36**, 1112-1120.

Lorey, F. & Smith, D. (1985). Bilirubin binding by the fractionated allelic components of heterozygous rhesus albumin. American Journal of Physical Anthropology, **68**, 169-171.

Melartin, L. & Blumberg, B. (1966). Albumin Naskapi. Science, 153, 1664-1666.

Pound, F. J. (1938). History and cultivation of the tonka bean with analysis of Trinidad, Venezuelan, and Brazilian samples. Tropical Agriculture, **15**, 28-32.

Smith, D.G. & Ahlfors, C. E. (1981). The albumin polymorphism and bilirubin binding in rhesus monkeys (*Macaca mulatta*). American Journal of Physical Anthropology, **54**, 37-41.

Tanis, R., Ferrel, R., Neel, J. V. & Morrow, M. (1974). Albumin Yan-2. Annals of Human Genetics, London, **38**, 179-190.

Wilding, C., Blumberg, B. S., Vessell, E. S. (1977). Reduced warfarin binding on albumin variants. Science, **194**, 991-994.

PART III

GENETIC DIVERSITY - APPLICATIONS AND PROBLEMS OF COMPLEX CHARACTERS

SIGNIFICANCE OF DERMATOGLYPHICS IN STUDIES OF POPULATION GENETIC VARIATION IN MAN

G. HAUSER

Histologisch-Embryologisches Institut der Universität Wien, Wien, Austria

INTRODUCTION

Of all human morphological traits, dermatoglyphics are practically unique in their stability in the face of changes in age and environment. Established in the third month of prenatal life, the intricate patterns of ridges on the fingers and toes remain constant (apart from growth in size) until death in old age, and only vary if there is severe trauma to the digit. The quantitative total ridge count is regarded as the most highly heritable of all human continuously varying characters, shown by comparison of the observed correlation coefficients between relatives of different degrees and those expected on a multifactorial hypothesis in which genetic determination is complete. This conclusion derives from the early work of Holt (1952, 1961, 1968) on families in Britain, but has recently also been demonstrated in tropical populations by Hreczko and Ray (1985). The modes of inheritance of individual pattern types, minutiae and other detailed features are still not fully established, but the numerous studies of twins, family members, and patients with chromosomal disorders show how strong is the genetic influence upon these.

There remains much to be done in mapping the distribution of dermatoglyphic features in tropical populations, though the map is far fuller today than it was twenty years ago. In digital features there appear to be frequency clines sweeping across the tropical areas, there appear to be differences between the major population groups inhabiting the tropics, and there appear to be quantum differences between small tropical populations ethnically different from their neighbours, as for example the differences between the Bushmen of the Kalahari or the pygmies of the Ituri Forest who are quite distinct from their later-coming Bantu neighbours (Hauser, 1977). While there are still large gaps in the detailed map of digital traits, information as regards the characters of the palms and especially the soles is notably sparse. But it is to draw attention to another aspect of dermatoglyphic work in the tropics that is the object of this paper, its application in microevolutionary studies to indicate and interpret similarities and differences among local populations.

POPULATION AFFINITIES

The Maya

One of the first demonstrations of what established dermatoglyphic methods can contribute to discerning biological distances and affinities between populations was carried out on tropical populations, using data on the Maya-speaking communities. Newman (1960) analysed the material collected by Stewart in 1947 and 1949 in surveys in four communities in south-west Guatemala, Soloma, Solola, Patzun, and Santa Clara La Laguna. The Soloma people speak Kanhobal, Solala and Patzun speak Cakchiquel, and Santa Clara Quiche. Using four simple variables (the index of pattern intensity, mainline index, thenar-first interdigital and hypothenar patterns), he showed that the two Cakchiquel and the Quiche samples from the Lake Atitlan basin in the South Highlands appeared to represent one biological population. Dermatoglyphic comparison suggested that this apparently single population was closely related to the Yucatan lowland Maya, principally represented by people from around Chichen Itza. Geographical distance precludes participation of the people of the two regions in the same breeding population, but the inference is that they belong to the same close grouping of such populations. These peoples have been separated for several centuries.

In contrast to the apparently close connection between these Lake Atitlan groups and the lowland Maya, the Soloma people in the North Highlands resembled dermatoglyphically those further north in Chiapas and together they stood out as quite a separate group of populations. The seemingly abrupt line of separation between the group in the North and those in the South Highlands follows the easterly extension of the Sierra Madre range and the southern escarpment of the Chuchumatanes Mountains. In addition to serving as a partial barrier to geneflow, this mountainous line also affords some cultural and linguistic separation. In their main dermatoglyphic features, the North Highland group appeared to have much closer relationships with Indians further north in Mexico than with their fellow Maya speakers.

This study suggests that dermatoglyphic features are essentially conservative, indicating biological affinities and differences that antedate present residence.

Bougainville

A second example comes from the work of Friedlaender (1975) on Bougainville. He surveyed eighteen villages, thirteen in the centre of the island covering three different language stocks, two from a Papuan language group further north, and three from one further south. He collected dermatoglyphic material from nearly fourteen hundred inhabitants, which he compared with the results of a serological survey on just over 2000, anthropometric data on 500, and dental data on some 900.

In the serological results, while there was considerable genetic variablity within each village, intervillage variation largely corresponded to language and geographic relationships. The anthropometric data showed clear evidence of secular change and change with age, but most of the variation occurring between villages reflected language group differences, bore little resemblance to the pattern predicted from current migration, and hence was to be interpreted as genetically determined. The dental data showed the same marked division between the northern Papuan-speaking and the southerners as did the anthropometry and blood polymorphisms. No such clear pattern was shown, however, in the dermatoglyphic results (palmar mainlines and ridge counts on each digit). Comparing these different lines of evidence by Gower's R^2, the correspondence between anthropometry and language was the closest, followed by anthropometry and blood polymorphisms, geography and blood polymorphisms, and geography and language. There was little correspondence of the dermatoglyphic results with any other group, and the closest correspondences were of male dermatoglyphics with migration, and female dermatoglyphics with blood groups. Somewhat similar results came from using Kendall's *tau*, and here the closest correspondence of dermatoglyphics was with migration and blood polymorphisms (males), and migration and geography (females). Friedlaender concluded that the more the phenotype varies among the villages, the closer the correspondence of the resulting pattern with the relationships predicted from migration, language and geographic information; but that dermatoglyphic variation is a relatively insensitive indicator of variation among these populations, as are other continuously varying traits such as skin colour, hair form and eye colour which are monotonously uniform over the area. He argued that the diversification which had characterised these villages over the past few centuries had not been sufficient to produce appreciable change in the dermatoglyphics, so that intervillage variation remains comparatively unimportant. In other words, dermatoglyphic features are once again shown to be conservative.

Venezuela

Rothhammer et al (1973) examined a similar problem, this time differentiation among villages of Yanomama Indians of southern Venezuela and northern Brazil. They showed considerable intervillage variation in most of the fifteen dermatoglyphic traits analysed, and again appreciable sex differences. The sample sizes, however, were small. The dermatoglyphic distances between the villages were subsequently examined (Neel et al, 1974) by the Mahalanobis multivariate statistic and the resulting matrix was compared with those for the same seven villages based on gene frequencies, anthropometry, and geographical distance. The correlations among the distance matrices (tau) are shown in the table, the closest being between gene frequencies and dermatoglyphics, and the least between anthropometry and dermatoglyphics. The values of tau are quite similar to those observed in the Bougainville study. Other assessments of similarity among the Yanomama were carried out, and the majority of comparisons involving dermatoglyphics indicated significant correspondence. Indeed, the matrix for the dermatoglyphics agreed almost as well with the other matrices as the latter did with each other. The finding, however, that the "best" network of dermatoglyphic distances appeared to differ somewhat from those of the other characters indicates the independence of the information that dermatoglyphic analysis provides. It gives a further dimension to such studies.

Table 1: Concordance of dermatoglyphic distances (τ)

Population	Dermatoglyphics with	Males	Females
Bougainville	Anthropometry	0.12	0.05
	Male dentition	0.17	0.01
	Female dentition	0.13	0.07
	Blood groups	0.29	0.22
	Geography	0.26	0.27
	Language	0.08	0.08
	Migration	0.35	0.29
Yanomama	Blood groups	0.34	
	Anthropometry	0.08	

New Guinea

In New Guinea, Giles obtained finger and palm prints from three distinct language groups, each represented by three villages or village clusters, and the large samples that he collected represented all of the existing people of each population. Using 35 dermatoglyphic traits which showed significant differences between the three main language groups (Froehlich & Giles, 1981a), discriminant functions were calculated, and from their contributions to the discriminant functions, a reduced set of fifteen variables was selected. The 95% confidence ellipses attaching to the discriminant group means showed considerable overlap. However, the first function discriminated between the Atsera and Waffa in that the former tended to have higher ridge counts and pattern complexities (except for the second and third fingers), lower mainline A, and higher atd angle, hypothenar, interdigital 3, and mainline B and C scores. The second function distinguished the Kukukuku from the other two, based on the combination of a high ridge count on the right thumb but with an antithetical lower pattern intensity. The application of discriminant analysis to these data was both efficient and efficacious. It suggested that the Melanesian/Papuan linguistic dichotomy (Atsera versus the other two) was reflected in distinct dermatoglyphic configuration. Two waves of Papuan migration were suggested, with reciprocal gene flow between the Atsera and Papuan peoples, and a possible migration route for the second Papuan migration into the New Guinea interior. It was concluded that dermatoglyphic characters, when the data are properly manipulated, can be seen to have a greater environmental and evolutionary stability than do serological and metrical traits, and that as biological markers of past population relationships, dermatoglyphics would offer new insights into the prehistory of New Guinea.

A further analysis (Froelich & Giles, 1981b) compared the serogenetic, anthropometric and dermatoglyphic population structures of the Markham valley villages. Rank order correlations between geographical, dermatoglyphic, anthropometric and serological distances showed the closest association of dermatoglyphics with geography, and significant associations with anthropometric and serological distances. Serological heterogeneity among the Waffa villages was attributed to random genetic drift; their greater homogeneity in dermatoglyphics suggested that these are more resistant to such stochastic processes. The authors concluded that the rates at which dermatoglyphic features change evolutionarily may be sufficiently slow for dermatoglyphics to be used for reconstruction of early population relationships.

DISCUSSION

Enough has been said to demonstrate the utility of dermatoglyphics in studies of local population affinities. This utility derives from their strong genetic determination, lifelong stability, absence of environmental influence, and apparent lack of appreciable differential selective advantage. Of course their informativeness depends upon the sizes of sample available, for when these are too small random error is likely to obscure underlying patterns. For them to be informative in population studies, sometimes the use of simple variables is sufficient, as in the Maya study, but more usually multivariate analytical procedures are required.

The results of the studies quoted are not strictly comparable since, for example, they differ in the number of dermatoglyphic characters included. Internally, it is not strictly justifiable to compare, as did two of the studies, results by discriminant function with those by principal components analysis. But they all agree that dermatoglyphic analysis gives a different perspective on local population structure and relationships, and provides a dimension that is largely independent of those of serological polymorphisms, anthropometry, and other variables. They all agree on the conservatism of dermatoglyphic features, that their rate of change is slower than the other variables considered. The most complex of the analyses quoted indeed goes so far as to claim that dermatoglyphic comparisons can be used to support hypotheses of migration and geneflow along likely geographical routes. Following this line of argument into speculation, it may well be possible to utilise their stability to enquire into their use in distinguishing between populations who have been environmentally stressed over long periods, and those that have not. But that is for another study.

REFERENCES

Friedlaender, J.S. (1975). Patterns of Human Variation. Cambridge (Mass.): Harvard University Press.

Froehlich, J.W. & Giles, E. (1981a). A multivariate approach to fingerprint variation in Papua New Guinea: implications for prehistory. American Journal of Physical Anthropology, **54**, 73-91.

Froehlich, J.W. & Giles, E. (1981b). A multivariate approach to fingerprint variation in Papua New Guinea: perspectives on the evolutionary stability of dermatoglyphic markers. American Journal of Physical Anthropology, **54**, 93-106.

Hauser, G. (1977). Das Hautleistensystem der afrikanischen Kleinwuchsigen. Mitteilungen der Anthropologischen Gesellschaft in Wien, **107**, 14-25.

Holt, S.B. (1952). Genetics of dermal ridges: inheritance of total ridge count. Annals of Eugenics, **17**, 140-161.

Holt, S.B. (1961). Dermatoglyphic patterns, In: G.A. Harrison (ed.), Genetical Variation in Human Populations, pp. 79-98. Oxford: Pergamon.

Holt, S.B. (1968). The Genetics of Dermal Ridges. Springfield: Thomas.

Hreczko, T. & Ray, A. (1985). An extended family study of ridge counts in two Indian populations. Human Biology, **57**, 289-302.

Neel, J.V., Rothhammer, F. & Lingoes, J.C. (1974). The genetic structure of a tribal population, the Yanomama Indians. X. Agreement between representations of village distances based on different sets of characteristics. American Journal of Human Genetics, **26**, 281-303.

Newman, M.T. (1960). Population analysis of finger and palm prints in Highland and Lowland Maya Indians. American Journal of Physical Anthropology, **18**, 45-58.

Rothhammer, F., Neel, J.V., La Rocha, F. & Sundling, G.Y. (1973). The genetic structure of a tribal population, the Yanomama Indians. VIII. Dermatoglyphic differences among villages. American Journal of Human Genetics. **25**: 152-166.

POLYMORPHISMS OF RED-GREEN VISION AMONG POPULATIONS OF THE TROPICS

A. ADAM

Everyman's University, Tel Aviv, Israel

INTRODUCTION

Though there is an apparent wealth of information on the occurrence of "colourblindness" among indigenous populations of the Tropics, there are serious shortcomings in much of it: many samples of well-defined populations are too small, large samples often combine subsamples of unspecified numbers of populations; in many studies, methods are used which underestimate the defect frequencies (Table 1) and others which cannot differentiate genuine red-green blindness from mild anomalous colour-vision, or indeed from underlying ophthalmological disorders.

This is particularly unfortunate, since such information is essential for any attempt to interpret this polymorphism in terms of evolutionary processes.

Table 1: Rates of "colourblindness" with and without anomaloscopy

		No. males tested	Literacy	% "colourblind" anomaloscopy -	+	Reference
Baganda	(1951)	537	+/-	1.9		Simon (1951)
	(1970)	718	+		3.9	Adam et al (1970)
Papuan-New Guineans	(1956)	4077	+/-	2.0		Mann & Turner (1956)
	(1969)	4820	+		4.7	Adam et al (1969a)
Thais and	(1963-7)	15000	+	4.0		Chotiprasit (1968)
Chinese	(1969)	2756	+		5.5	Adam et al (1969b)
Aymara	(1966)	140	+/-	3.6	6.4	Cruz-Coke & Barrera (1966)

Needless to say, the Tropics and its multitude of peoples provide ideal opportunities for investigations of the outcome of genetic drift, migration, selection or relaxed selection, factors likely to be involved in the maintenance, or change, of this polymorphism. For the Tropics are the home of virtually all types of human societies, from small hunter-gatherer groups to some of the largest and oldest civilisations on earth, and the changes that are occurring there are particularly relevant to the hypothesis that the colour defect gene frequency variation seen today is the outcome of relaxation of selection pressures exerted in traditional life styles (Post, 1971).

BIOLOGY OF RED-GREEN COLOURBLINDNESS

Red-green vision is governed by two adjacent X-linked, multiple allelic loci, protan and deutan. Only two of the mutant alleles, one at each locus, cause red-green blindness. In the retinal cones of normal people (trichromats) the pigment is of 3 types - red-absorbing, green-absorbing and blue-absorbing. In affected individuals the gene for red or green blindness leads to the absence of either the red-absorbing or the green-absorbing pigment from the retinal cones, while the blue-absorbing pigment is intact. Thus affected people (dichromats) are unable to distinguish any difference in hue over the whole red-orange-yellow-green region of the spectrum, and they confuse many other, non-spectral colours as well. The first and severest type of red-green blindness, protanopia, was described by Dalton at the end of the 18th century, and the second, deuteranopia, has been known since 1837.

In 1881 Lord Rayleigh discovered "an interesting peculiarity of colour vision, quite distinct from colour blindness" (Rayleigh, 1881): he described the two most common aberrations of red-green vision, that are now called "simple" protanomaly and deuteranomaly. These people have 3 pigments in their cones, and therefore they, like normals, are trichomats. The common anomalous phenotypes are due to two mutant alleles at the respective loci, each being recessive to normal red-green vision and dominant over the mutations responsible for the more severe "anopias" and other rare extreme anomalies. Rayleigh's discovery was made while he was experimenting with an instrument that he constructed whereby red and green lights could be mixed in all proportions, and the resulting colours compared to an adjacent yellow with adjustable brightness. The anomaloscope is an indispensable tool for research and diagnosis of the various defects of red-green vision. The dramatic contrasts between dichromatic red-green blindness and anomalous trichromatic vision is illustrated in Figure 1: whereas the red-green blind dichromats accept all mixtures of red and green as perfectly matching the standard yellow, the red and green anomalous trichromats accept only one characteristic mixture, each removed in a different direction from the normal mid-matching point. Rayleigh emphasised that the colour vision of the latter "is defective only in the sense that it differs from that of the majority" (Rayleigh, 1881): in contrast to the red-green blind, they not only distinguish easily red from green, but also appreciate small colour differences in between.

It is not surprising that genuine colourblindness, and particularly protanopia, is usually noticed early in childhood, both by the affected person and by his close relatives. In contrast, a great many anomalous trichromats, andparticularly deuteranomals, are not aware of their mild peculiarity, and are

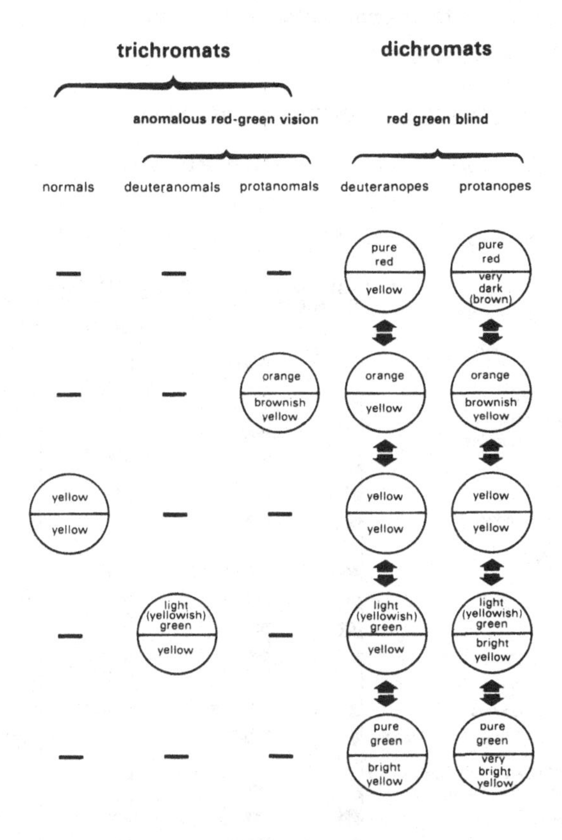

Figure 1: Mixtures of red and green lights (upper halves of circles) accepted by the five most common types of observers as perfect matches to a standard yellow of adjustable brightness (bottom halves of circles). The five depicted matches of protanopes and deuteranopes are representatives of infinite number of matches they accept, over the full range of red-green mixtures.

astonished to discover that they fail a screening test for colourvision, e.g. the Ishihara. Such screening tests have their drawbacks: the Ishihara test groups the anomalous indiscriminately with the colourblind (Pickford, 1951), and is inadequate for quantitative diagnosis (Crone, 1961). Unfortunately it is on such tests that the majority of population studies of this polymorphism, particularly in the Tropics, have been based.

It has been argued (e.g. Pickford, 1963; Adam, 1969) that the contrasting colour discriminating performances between the anomalous and the colourblind affects their behaviour in natural environments. Therefore there is likely to be differential pressure on mutant phenotypes. To assess this, it is essential to have precise diagnoses and estimates of allele frequencies and these can only be obtained by the use of portable anomaloscopes (Kalmus et al, 1964; Adam et al, 1969; Pickford & Lakowski, 1960).

GENE FREQUENCIES IN THE TROPICS

All available, anomaloscopically-derived estimates for the frequencies of the four common mutant colour-vision alleles among indigenous tropical populations, are condensed into Figure 2. Most of the samples consist of subgroups,

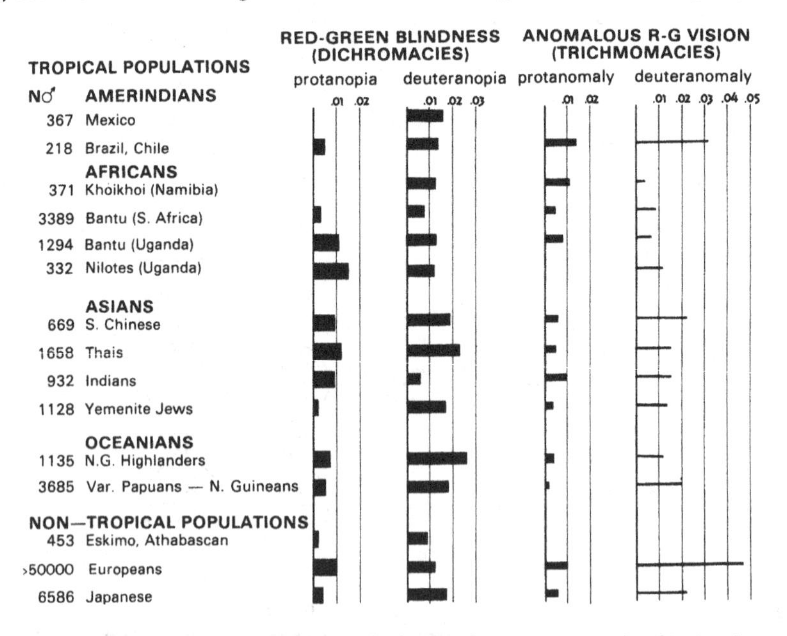

Figure 2: Available estimates of allele frequencies obtained through anomaloscopy (Adam, 1985).

often displaying wide variation in the overall rates of defects, and in allele frequencies. A few tentative statements may be made regarding the frequencies themselves:

1. Protanopia is probably very rare among Amerindians and a few other tropical groups of all continents.

2. Deuteranopia is apparently not rare in any of the tropical population groups examined; it seems to be even more frequent in some of the so-called "primitive" populations than among some long-established civilisations (Adam et al, 1969a).

3. The combined frequencies of the two dichromatic defects, i.e. of genuine red-green blindness, in the Tropics and elsewhere, vary generally within the narrow range of about 1.0-3.5%, with few exceptions. Both the lower and the higher frequencies within this range occur in primitive as well as in advanced societies.

4. The two types of anomalous red-green vision are not particularly frequent in any of the tropical population groups. The combined frequency of the anomalies is usually similar, and often lower, than that of the anopias. Such a proportion between anomalies and anopias is strikingly different from that found among all Europeans. Deuteranomaly at a very high frequency is a constant characteristic of the Europeans, but it is rare elsewhere and has been found so far only in two tropical populations, both in Papua.

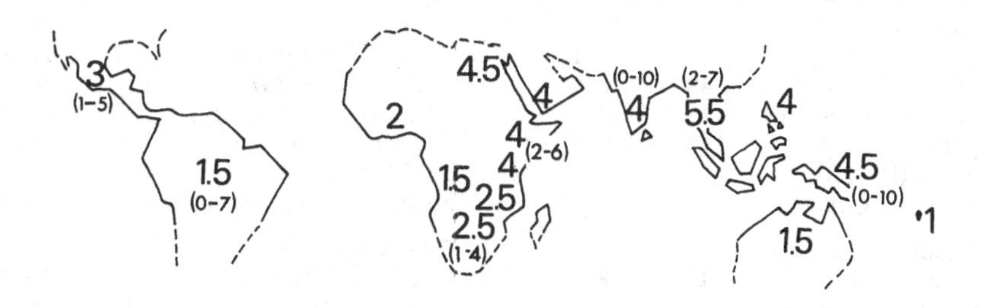

Figure 3: Average percentages and range of overall red-green defects among males in indigenous populations of the Tropics. Major references for America: Kalmus et al (1964), Salzano (1980), Cruz-Coke & Barrera (1966); for Africa: Adam (1981); for India: Dutta (1966); for Thailand: Adam et al (1969b); for Oceania: Mann & Turner (1956), Adam et al (1969a).

On the whole, the impression from studies in the Tropics (Fig. 3) is that colourblindness is indeed very rare in some populations (Table 2) and quite common in others (Table 3). This heterogeneity occurs in both "primitive" populations of all continents and old civilisations. Furthermore, the whole range of frequencies may be found in subsamples of one group, like the highest class of traditional Hindu society, the Brahmins, as in Table 4. But so far there is little experimental work to enquire into the hypotheses of selection, and there is moreover no proof that alleles which are apparently the same in different populations are in fact identical in their molecular structure. Both of these require urgent attention.

Table 2: Tropical populations with apparently monomorphic, normal red-green vision

Many South Amerindian tribes
Pygmies (Central Africa)
Central San (Bushmen, South Africa)
A few Brahmin and other Hindu groups, Assam tribes, etc.
Several Oceanian tribes

Major references are those of Figures 2 and 3.

Table 3: Highest frequencies of overall red-green defects among indigenous male populations of the tropics

	No. tested	% colour-blind		No. tested	% colour-blind
Amerindians			**Asians**		
Jivaro (Equador)	183	7.1	Ismailis (in Uganda)	261	13.8
Kaingang (Brazil)	43	7.0	Toda (S.E. India)	823	10.2
Yanomama (Brazil)	173	6.9	Vadanagara (Bombay)	100	10.0
Aymara (N. Chile)	146	6.4	Audichya (Bombay)	100	9.0
Caraja (Brazil)	35	5.7	Koli (Gujarat)	67	9.0
Tlaxcala (Mexico)	92	5.4	Sudra (Andhra Pradesh)	180	8.9
Africans			**Oceanians**		
N. San (Bushmen)	46	8.7	Karkar Island (N.G.)	246	10.2
Amhara (Ethiopia)	142	6.3	Roro (Papua)	139	9.3
Rendille (Kenya)	56	5.4	Trobriand Islands	123	6.5
Nyankole (Uganda)	190	5.3	Motu (Papua)	437	6.4
Nilotes (Uganda)	185	4.9	Toaripi (Papua)	215	6.0
Billen (Ethiopia)	105	4.8	Chimbu (Highlands, N.G.)	220	5.9

Major references are those of Figures 2 and 3.

Table 4: Colourblindness among various male Brahmins in India (Dutta, 1966)

	No. tested	% Colourblind
Desath (Madras)	153	0
Vandanagara (Varanasi)	86	1
Madhya Pradesh	243	3
Orissa	186	4
Koknasth (Bombay)	100	5
Andhra Pradesh	83	6
Rarhi (Orissa)	217	7
Audichya (Bombay)	100	9
Vadanagara (Bombay)	100	10

REFERENCES

Adam, A. (1969). A further query on color blindness and natural selection. Social Biology, **16**, 197-202.

Adam, A. (1981). Colorblindness in Africa. Met.Ped.Ophthal. **5**, 181-185.

Adam, A. (1985). Colorblindness in man: a call for re-examination of the selection-relaxation theory, In: Y. R. Ahuja & J. V. Neel (eds.), Genetic Microdifferentiation in Human and Other Populations, pp. 181-194. Indian Anthropological Association, Delhi.

Adam, A., Wood, W. B., Symons, C. P., Ord, I. G. & Smith, J. (1969a). Deficiencies of red-green vision among some populations of mainland Papua and New Guinea. Papua New Guinea Medical Journal, **12**, 113-120.

Adam, A., Puenpatom, M., Davivongs, V. & Wangspa, S. (1969b). Anomaloscopic diagnoses of red-green blindness among Thais and Chinese. Human Heredity, **19**, 509-513.

Adam, A., Mwesigye, E. & Tabani, E. (1970). Ugandan colorblinds revisited. American Journal of Physical Anthropology, **32**, 59-64.

Chotiprasit, S. S. (1963). Colourblindness in Thailand. EENT News 1: 24-31.

Crone, R. A. (1961). Quantitative diagnosis of defective color vision. A comparative evaluation of the Ishihara test, the Farnsworth dichotomous test and the Hardy-Rand-Rittler polychromatic plates. American Journal of Ophthalmology, **51**, 298-305.

Cruz-Coke, R. & Barrera, R. (1966). Color blindness among Aymara in Chile. American Journal of Physical Anthropologists, **31**, 229-230.

Dutta, P. C. (1966). A review of the inherited defective colour-vision variability and selection relaxation among the Indians. Acta Genetica (Basel), **16**, 327-339.

Kalmus, H., DeGaray, A. L., Rodarte, U. & Cobo, L. (1964). The frequency of PTC tasting, hard ear wax, colour blindness and other genetical characters in urban and rural Mexican populations. Human Biology, **36**, 134-145.

Mann, I. & Turner, C. (1956). Colorvision in native races of Australia. American Journal of Ophthalmology, **41**, 797-800.

Pickford, R. W. (1951). Individual differences in colour vision, p. 236. London: Routledge and Kegan Paul.

Pickford, R. W. (1963). Natural selection and colour blindness. Eugenics Review, **55**, 97-101.

Pickford, R. W. & Lakowski, R. (1960). The Pickford-Nicolson anomaloscope. British Journal of Physiological Optics, **17**, 131-150.

Post, R. H. (1971). Possible cases of relaxed selection in civilized population. Humangenetik, **13**, 253-284.

Rayleigh, Lord (1881). Experiments on colour. Nature, **25**, 64-66.

Salzano, F. M. (1980). New studies on the color vision of Brazilian Indians. Brazilian Journal of Genetics, **3**, 317-327.

Simon, K. (1951). Colour vision of Baganda Africans. East African Medical Journal, **28**, 75-79.

REDUCTIONS IN BODY SIZE AND THE PRESERVATION OF GENETIC VARIABILITY IN TROPICAL POPULATIONS

W. A. STINI

Department of Anthropology,
University of Arizona, Tucson, U.S.A.

INTRODUCTION

Among the human populations of the world, those of the tropics are currently experiencing the highest mortality and fertility rates (Table 1) so that population turnover is generally rapid. The characteristic combination of high mortalities and high fertilities, so different from the situation in Western countries, maximizes the opportunity for natural selection influence. For under such circumstances, strong selection against specific phenotypic characteristics should produce altered gene frequencies. Where strong and persistent selection continues over many generations, it would be anticipated that genetic diversity would be diminished.

Comparison of mortality statistics for men and women in developing countries with those observed in the industrialised nations shows further differences. For instance in 1978, male and female life expectancies in both U.S.A. and Canada were approximately 69 and 76 years respectively, and 73 and 79 years in Iceland (United Nations, 1978), quite a large difference between the sexes. In developing countries (Table 1, columns 2 and 3) there is a very small difference in the life expectancies of males and females, in a few cases even a reversal of the sex trend seen in industrialised areas. A clue to the reason can be found in the fertility and mortality statistics (columns 4 and 7). In developing countries, where pre- and peri-natal care of both the infant and mother are often minimal, the children are at risk early in life and the mother in early adulthood. For instance, deaths per thousand in the Western industrial countries average about 9 compared to more than 25 in Upper Volta. The differences in infant mortality are even more striking, with an average rate of about 10 per thousand in the Western countries compared to 182 in Upper Volta and 159 in Liberia.

What the mortality statistics illustrate is that in many of the tropical

Table 1: Life expectancies at birth, birthrates, fertility rates and infant mortalities in nations where male life expectancy equals or exceeds female life expectancy at birth

	Life expectancy		Births	Fertility	Deaths	Infant mortality
	Male	Female	/1000		/1000	/1000
Upper Volta	32.10	31.10	48.5	197.0	25.8	182.0
Bangladesh	35.80	35.80	49.5	231.7	2.8	S
Nigeria	37.20	36.70	49.3	217.8	S	
India	41.89	40.55	34.5	136.7	14.4	122.0
Kampuchea	44.20	43.30	46.7	143.1	19.0	127.0
Liberia	45.80	44.00	49.8	161.2	20.9	159.2
Sabah Malaysia	48.79	45.53	35.0	179.4	14.4	31.6
Indonesia	47.50	47.50	42.9	175.7	16.9	125.0
Pakistan	53.72	48.80	36.0	174.8	12.0	124.0
Jordan	52.60	52.00	47.6	206.6	14.7	36.3
Industrial nations of high life expectancy						
Canada	69.34	76.36	15.8	61.6	7.2	15.0
U.S.A.	68.70	76.50	14.7	58.5	8.9	15.1
Denmark	71.10	76.80	12.9	61.3	10.7	10.3
France	69.00	76.90	13.6	72.0	10.5	10.4
The Netherlands	71.20	77.20	12.9	53.9	8.3	10.5
Japan	72.15	77.35	16.4	62.6	6.3	9.3
Sweden	72.09	77.65	11.9	56.4	11.0	8.7
Norway	71.50	77.83	13.3	64.1	9.9	11.1
Iceland	73.00	79.20	19.4	81.9	6.9	11.7

Source: Statistical Yearbook 1977, United Nations Department of Economic and Social Affairs, New York, 1978.

Table 2: Mortality rates (per 1000) among children under age 5 years in Maltab, Bangladesh
(adapted from Chen, Rahman & Sarder, 1980)

Cause of death	Infant mortality (birth to 1 yr)	Mortality 1-4 yrs
Diarrhoea	19.6	15.1
Watery	16.8	7.4
Dysentery	2.8	7.6
Tetanus	37.4	0.6
Measles	3.1	4.5
Fever	7.3	2.9
Respiratory	10.4	1.6
Drowning	0.6	2.2
Skin diseases	1.9	0.4
Others	62.2	7.0

regions of the world death occurs early, during or even before the reproductive years. As a consequence of this early mortality, the opportunity for natural selection to remove phenotypes lacking effective adaptive capabilities to survive in the tropical regions of the world is high. This kind of natural selection has probably been the rule rather than the exception through most of human history in most of the world and, in consequence, it is not unlikely that most human populations share certain adaptive strategies as a result of thousands of generations of such selection. The tropical regions of the world today show a continuation of the pattern, with disease and malnutrition forming a combination of stresses that will tend to lead to early rather than late mortality.

DATA FROM CONTEMPORARY BANGLADESH

Even today, mortality rates in Bangladesh approximate 25% between birth and age 5. More than half of the total deaths in these communities each year occur among the one-sixth of the population under the age of five years. As a result, life expectancy at birth in Bangladesh is fifty years, whereas a child who survives to age one has a life expectancy of 57 years and those who are still alive at age five can expect to live to an average age of 63 (International Centre for Diarrhoeal Disease Research, Bangladesh, Newsletter November - December, 1984).

According to an extensive three-year survey of causes of death among 260000 people living in 228 rural Bangladesh villages, mortality rates for the first year of life in these villages averaged 142.6 per 1000 live births (Chen et al, 1980). The single most important cause of death, accounting for 26% of the total was tetanus during the first four weeks of life. Diarrhoea, fevers and respiratory disease accounted for an additional 26%, with watery diarrhoea alone being cited as the cause of death in 11.8% of these infant mortalities. Over 43% of infant deaths, largely those of neonates, were not categorised, with a variety of factors such as prematurity, low birth weight, obstetric difficulties, birth injuries and infections, asphyxia and congenital anomalies all contributing to the mortality in this early stage of life.

If a child survives the hazards of the first years of life, chances of survival improve substantially, with a mortality rate of 34.3 per thousand in the age group 1-4 years. During this stage of childhood, about 44% of mortalities were directly attributed to either watery diarrhoea of dysentary. Measles was responsible for about 13%, fever and respiratory diseases about 13% and, in this flood-prone area, drowning over 6%.

In these Bangladesh villages, a distinct seasonal pattern can be discerned in causes of mortality. For instance during the months of September through December, when most births occurred, tetanus deaths were most prevalent, while the dry months of March and April were the time of high mortality from measles. The winter months of December to March were a time of many deaths from respiratory diseases, a pattern familiar throughout Asia. The time of monsoon flooding was the time of high mortality from watery diarrhoea of several forms including cholera. The months of July through September were also the time of many deaths by drowning.

There is a clear difference between the numbers of early male and female deaths. Among neonates, mortality was much higher among male children while female mortality was higher thereafter. The greatest disparity was seen among the one-year olds where female mortality exceeded male for all causes except tetanus (a neonatal condition, for the most part) and drowning. The researchers surmised that in this population, socio-cultural factors, specifically the preference for male children with the associated low status of women, led to preferential care for males including biased food allocations and health care.

Major risk factors that were identified in these Bangladesh villages were socioeconomic and nutritional status. Crowded and substandard housing concomitant with low socioeconomic status was shown to be a predictor of mortality rates roughly double those observed among better-housed children. Also, mortality rates three times those seen in the better-nourished segment of the population were found in children whose weight-for-age was below 65% and weight-for-height below 70% of Harvard standards. In general, the relationship between high mortality rates, housing and poor nutritional status was consistently close; an observation that is not surprising when the common influence of low socioeconomic status is taken into account. There is, of course, a certain amount of ambiguity in the interpretation of mortality statistics in populations where simultaneous occurrence of two or more possible causes of death is often the rule. For instance, measles and diarrhoea are frequently seen in the same child and when their combined effects lead to death, the ultimate cause may be difficult to determine, with the attribution becoming more a matter of judgment than of certainty.

TROPICAL POPULATIONS AS A FOCUS OF CONTEMPORARY NATURAL SELECTION

The high mortality rates observed in tropical countries are, in large part, attributable to the combined effects of disease and nutritional deficiencies. Both are subject to seasonal variation and both may occur as endemic, epidemic or pandemic. Each is implicated in the etiology of the other, as the Bangladesh example illustrated. In tropical regions, where year-round survival of pathogens is the general rule, the effects of disease are frequently modulated by endemic undernutrition of deficiencies of specific nutrients such as essential amino acids, vitamin A or one of the B-vitamins. Universal exposure to a wide range of pathogens early in life means that a premium is placed on the ability to mount a suitable immune response (Good, 1977). Thus, the ability to maintain growth, while at the same time diverting the body's resources to support the immune response at the expense of growth, is of central importance in the face of the prevailing constellation of stressors in the tropics. Thus, body size, especially lean body mass, is significantly reduced (Table 3). The size reduction is more pronounced in males than in females (Hiernaux, 1968; Stini, 1972a, 1972b, 1975), and the net effect is a reduction in sexual dimorphism for body size.

Can this observed reduction in size dimorphism in human populations be related to the maintenance of genetic diversity under the conditions prevailing in the tropics? Closer examination of some of the factors involved will, I believe, give convincing evidence that it can. The differences in metabolic requirements of male and female (Table 4) means that the relative energy cost of supporting males and females in a well-fed population is different. The amount of energy that is required to support a well-fed male is greater than that necessary to support one who has undergone endemic nutritional inadequacy through the critical phases of growth and development (Table 5). Although it is not valid to extrapolate directly from such observations to entire populations, a reasonable estimate would be that the same supply of energy would support, in such restricted circumstances, a population of males 10% greater than would be possible if the full growth potential was achieved by all. Taken alone, this factor would be instrumental in the preservation of more genotypes than would otherwise occur, and as a consequence, must be considered an important mechanism for the preservation of genetic diversity in tropical populations.

However important size reduction in males may be in the preservation of genetic diversity in tropical populations, its significance is considerably enhanced by the difference in male and female responses to endemic nutritional

Table 3: Mean upper arm muscle cross-section area of adults of both sexes: U.S.A. and Colombian values

	Colombia		U.S.A.		Colombia % of U.S.A.	
	Arm muscle area (mm²)	Body wt. (kg)	Arm muscle area (mm²)	Body wt. (kg)	Muscle area	Body wt.
Male	4267	60	6464	70	66	85
Female	3450	51	4272	58	81	88
Female % of male	81	85	66	83		

From: Stini (1975b)

Table 4: Energy expenditures (kcal/min) of mature reference man and woman of various activity levels

	Man 70 kg			Woman 58 kg	
	Time (hrs)	Rate (kcal/min)	Total (kcal)	Rate (kcal/min)	Total (kcal)
Sleeping	8	1.0-1.2	540	0.9-1.1	440
Very light	12	2.5	1300	2.0	900
Light	3	2.5-4.9	600	2.0-3.9	450
Heavy	1	7.5-12.0	300	6.0-10.0	240
Total	24		2740		2030

From: National Research (1980)

Table 5: Comparison of caloric requirements of a 70 kg U.S. male with those of a 60 kg Colombian male at similar activity levels

	U.S.A.	Colombia
Mean body weight (kg)	70	60
Caloric costs (kcal)		
Resting (8 hours)	570	480
Very light activity (8 hours)	630	540
Light labour (8 hours)	2624	1392
Moderate labour (2 hours)	602	516
Total 3426	2918	

From: Stini (1975b)

Table 6: Relationship of adipose tissue to body weight in adult humans

	Ratio	
	Adipose tissue mass: Body mass	
	Male	Female
25	0.14	0.26
40	0.22	0.32
55	0.25	0.38

From: Masoro, E. (1977)

inadequacies and disease exposure. Not only do more male neonates die in the perinatal period but, in the absence of culturally mediated factors (some of which affect the Bangladesh figures), underfed female children grow more similarly to well-fed ones and, although their sexual maturation is delayed, it is delayed to a lesser extent than that of males. The net effect is the maximisation of effective population size, as reflected in the number of reproducing females, for the energy required from a limited resource base. In addition, as can be seen in table 6, not only does the body composition of females determine their different metabolic requirements, it also ensures adequate reserves to support pregnancy and lactation despite cyclical food shortages (Masoro, 1977; Bourges et al, 1977; Martinez & Chavez, 1971).

FEMALE BODY COMPOSITION AND ITS REPRODUCTIVE SIGNIFICANCE

In the human species, the burden of reproduction falls most heavily on the female. From the standpoint of the physiological demands associated with bearing and rearing young, females are the sex most committed to the perpetuation of the species and are, by the same token, the determinants of the reproductive success of the population. Nevertheless, there are few if any societies where women do not in addition play an important role in the subsistence activities of the population. This is certainly true of the agrarian societies of the tropics. The demands of pregnancy and lactation often occur at the same time that heavy, physical labour must also be undertaken by women. Moreover, in areas where rainy and dry seasons punctuate the year with periods of fast and famine, a nine-month pregnancy followed by six months of breast feeding will guarantee that women in the reproductive stage of life will experience extraordinary demands on their reserves not only of energy, fats and amino acids but also of other body constituents. Therefore, the need for nutrient reserves may be acute for women in a way never experienced by men. Being as closely linked to reproductive success as it is, the ability to maintain and utilise reserves of energy, amino acids, fats, vitamins and minerals is a focus of strong selective pressure (Stini et al, 1980). Early abortion may be traumatic for a woman, but in terms of effect on her total reproduction, except in those cases where it causes lasting damage to her reproductive capability, it is a far less serious loss than loss of a newborn or, worse yet, the loss of an infant after six months of breastfeeding.

Because of the increasing investment of the mother and, for that matter, the entire community, in the survival of an infant after birth, evolution has continued to favour mechanisms whereby infant survival is assured even though maternal nutrition is inadequate. The major mechanism for support of the human infant during the early months of life is, of course, through breastfeeding. Human milk composition reflects the nutritional needs of the newborn in highly specific ways.

The need for reserves of fats and carbohydrates are, for adaptive reasons, greater in females than in males. Thus, in humans as in many other species, sexual dimorphism can be considered the product of disruptive selection, with the survival of the species being ensured by the attainment of different body composition in the male and female adult. Not only does body composition itself differ, with males pursuing a developmental path emphasizing increases in lean body mass while females acquire substantial amounts of fat, but the timing of the completion of the maturation process itself differs substantially with females maintaining an approximate 10% lead throughout. Also, males are more responsive to environmental influences than are females, both in terms of their size and body composition and in terms of the delay in attainment of maturity. Viewed as an adaptive strategy, such delays are more tolerable in males than in females. In an environment where many stresses are encountered, mortality is high and life expectancy short, it is important that females be capable of reproducing while still young. Environmentally-induced delays in female sexual maturity would severely limit the number of reproducing females available in a population as a proportion of the total number of individuals to be supported on the resources available. Thus, while delays in male maturation might be tolerated if males can function in some productive role while still immature and, ultimately, at reduced body size, the reproductive success of the population would be threatened by a similar alteration in the female growth and development process. In other words, there are good adaptive reasons for the canalisation of female growth and development, and these reasons have a sufficiently long history to have caused integral changes in part of the human genome. However, the greater responsiveness to environmental factors is seen most clearly in areas where nutritional stress is frequently encountered. The populations of the humid tropics are at present those most often so stressed and are therefore of great interest to the student of human adaptability.

CONCLUSIONS

The evidence that physiological adaptations involving the exploitation of phenotypic plasticity allow the maintenance of genetic diversity in contemporary tropical populations is indirect but abundant. However, many questions remain concerning the mechanisms that make such adjustments possible. Perhaps the most enigmatic aspect of the process is the nature of the metabolic adjustments that permit undernourished women to support repeated pregnancies and lactation in the absence of adequate protein and energy supplements. That significant changes in metabolism occur is empirically demonstrable, but the endocrinological factors and intracellular adaptations that effect them are poorly understood. As the events at the cellular level that undoubtedly underlie these adaptations become better understood, it may be possible to identify directly genetic determinants of the mechanisms that facilitate the maintenance of genetic diversity in tropical populations.

REFERENCES

Bourges, H., Martinez, C. & Chavez, A. (1977). Effect of dietary supplements on nutrient content of milk from mothers in a rural Mexican town. Paper presented at the Western Hemisphere Nutrition Congress V. Quebec, Canada, August, 1977.

Chen, L. C., Rahman, M. & Sarder, A. (1980). Epidemiology and causes of death among children in a rural area of Bangladesh. International Journal of Epidemiology, **931**, 25-33.

Good, R. A. (1977). Biology of the cell-mediated immune response - a review. In: R.M. Suskind (ed.), Malnutrition and the Immune Response. New York: Raven.

Hiernaux, J. (1968). Variabilite du dimorphisme sexual de la stature en Afrique Sub-Saharienne et en Europe. In: Sonderduck aus Anthropologie und Humangenetik. Stuttgart: Gustav Fischer Verlag.

Martinez, C. & Chavez, A. (1971). Nutrition and development in infants in poor rural areas. 1. Consumption of mother's milk by infants. Nutr.Rep. Internat., **16**, 356-359.

Masoro, E. J. (1975). Other physiologic changes with age. In: A.M. Ostfeld (ed.), Epidemiology of Aging. DHEW Publication no. NIH 77-711, Bethesda, pp. 137-155.

National Research Council (1980). Recommended Dietary Allowances, 9th ed. National Academy of Sciences, Washington, D.C.

Stini, W. A. (1972a). Malnutrition, body size and proportion. Ecol. Food Nutr., **1**, 126-132.

Stini, W. A. (1972b). Reduced sexual dimorphism in upper arm muscle circumference associated with protein-deficient diet in a South American population. American Journal of Physical Anthropology, **36**, 341-352.

Stini, W. A. (1975). Adaptive strategies of human populations under nutritional stress. In: E. Watts, F.E. Johnston and G.W. Lasker (eds.), Biosocial Interrelations in Population Adaptation, pp. 19-41. The Hague: Mouton.

Stini, W. A., Weber, C. W., Kemberling, S. R. & Vaughan, L. A. (1980). Lean tissue growth and disease susceptibility in bottle-fed versus breast-fed infants. In: L. Green (ed.), Social and Biological Predictors of Nutritional Status, Growth and Neurological Development. New York: Academic Press.

United Nations/World Health Organization (1978). Statistical Yearbook, 1977. New York: United Nations.

GENETIC ISOLATES AND THE SEARCH FOR CAUSAL MECHANISMS OF DISEASE

R. M. GARRUTO

Laboratory of Central Nervous System Studies, National Institutes of Health, Bethesda, Maryland, U S A

INTRODUCTION

Genetic isolates provide excellent opportunities to study normal and pathological processes. Such isolates are characterised by inbreeding, a geographically and culturally restricted territory, a "simplified" ecology and fixed habitat, a close association with flora and fauna, and unique social and behavioural patterns often inextricably tied to a particular disease expression or unusual epidemiologic pattern. Indeed, biomedical problems are often more appropriately studied in isolates than in large cosmopolitan communities, where numerous genetic, cultural and environmental variables complicate the analysis. The study of these "natural experiments" has already resulted in major biomedical discoveries.

The natural history of disease in isolated groups represents a system of dynamic interaction. Factors affecting the expression of disease in such populations include the degree of isolation and contact, degree of genetic relatedness, demographic characteristics including population size, density, and age and sex distribution, group mobility, ecosystem stability, uniformity of food sources, physical environmental stress, close physical proximity during work and play, culturally-specific hygienic practices, and the degree of natural resistance and differential genetic susceptibility. A model of such a system is suggested in Figure 1, where the isolate is schematically represented by a genetic homogeneity, a geographically and culturally restricted territory, nutritional, cultural and biological homogeneity, a "simplified" ecology and fixed habitat, a close association with flora and fauna, and by unique biological and/or behavioural patterns or specific disease phenomenon.

Many genetic or traditional isolates are located in tropical regions of the world. This paper presents examples from our research on tropical Pacific populations, to show how they have helped solve etiological and epidemiological

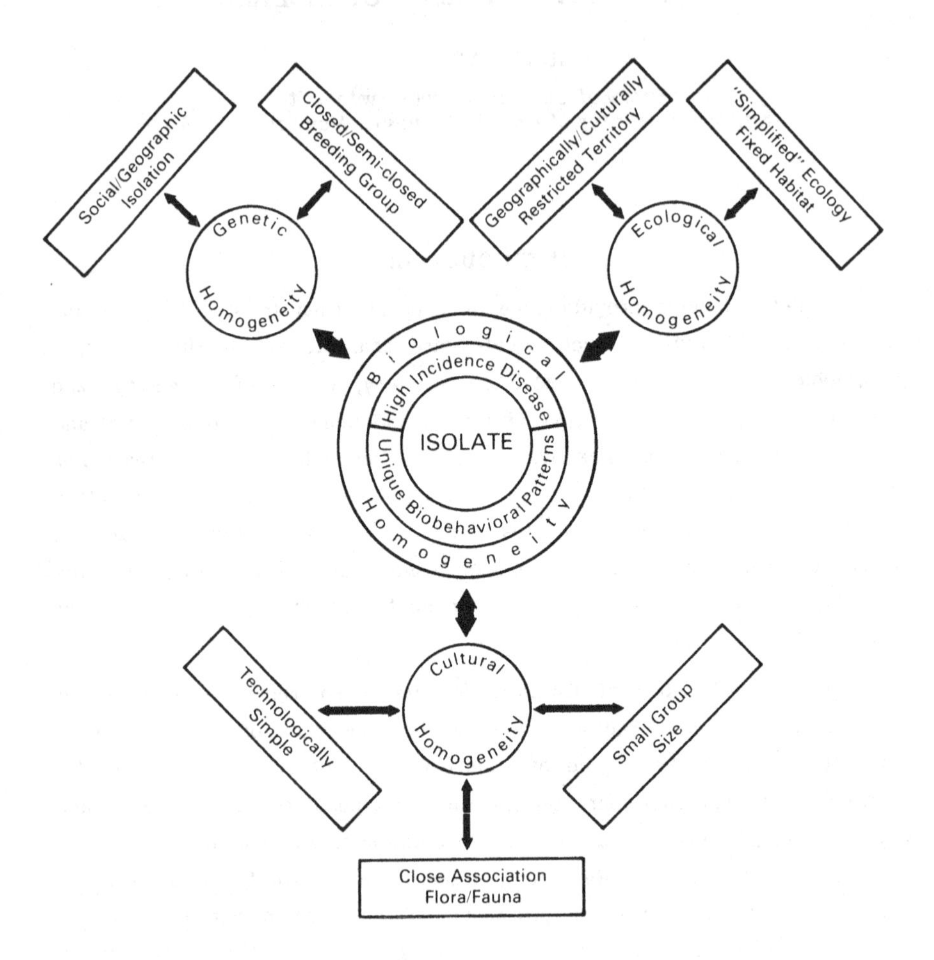

Figure 1: Natural experimental model for the study of health and disease in isolated human groups.

problems of wider medical significance. In particular, our use of such isolates to elucidate the etiology and pathogenesis of neurological disease in high incidence foci has been highly successful. Examples include the study of amyotrophic lateral sclerosis (ALS) and parkinsonism-dementia (PD) in the western Pacific, hyperendemic goiter and cretinism in West New Guinea, and the well known discovery of kuru among the Fore people of Papua New Guinea.

TROPICAL PACIFIC POPULATIONS: PARADIGMS FOR THE STUDY OF LATE ONSET NEUROLOGICAL DISORDERS

During the past 30 years, unusually high-incidence foci of amyotrophic lateral sclerosis (ALS) and parkinsonian syndromes have been recognised in, and later investigated among, three geographically and genetically distinct populations in the western Pacific region: the Chamorros of the Mariana Islands, the Auyu and Jakai people of West New Guinea, and the Japanese of the Hobara and Kozagawa districts on the Kii Peninsula of Honshu Island (Figure 2) (Garruto et al, 1985; Kurland & Mulder, 1954; Reed & Brody, 1975; Shiraki & Yase, 1975).

A disease of the motor neurons of the brain and spinal cord, ALS is characterised by muscle weakness, progressive atrophy of skeletal muscles, and paralysis. By comparison, parkinsonism is characterised by slowness of voluntary motor activity, muscular rigidity, tremor, and mask facies, and, in these foci, is usually accompanied by progressive mental deterioration. Both disorders are late-onset, uniformly fatal, and of unknown origin. The hyperendemic foci of ALS and parkinsonism, all located in the western Pacific, can be regarded as natural experimental models of late-onset chronic degenerative disease that occur in different cultures, in different ecological zones and among genetically divergent populations.

In the Kii Peninsula, two separate foci of ALS and parkinsonism have been discovered: in Kozagawa (population, 6200), and in Hobara (population, 2100) (Shiraki & Yase, 1975). The original incidence in the combined foci was more than 50 times that for the rest of Japan. These foci are 200 km apart and are located in remote, mountainous regions with farming and lumbering the main occupational activities.

The most remote focus of ALS and parkinsonism is found among the Auyu and Jakai people living on the southern coastal plains of West New Guinea. This focus, first described by Gajdusek, has been intensively studied during the past decade (Gajdusek, 1963; Gajdusek & Salazar, 1982). The overall crude average annual incidence for ALS alone is 147 per 100,000 population, almost 150 times

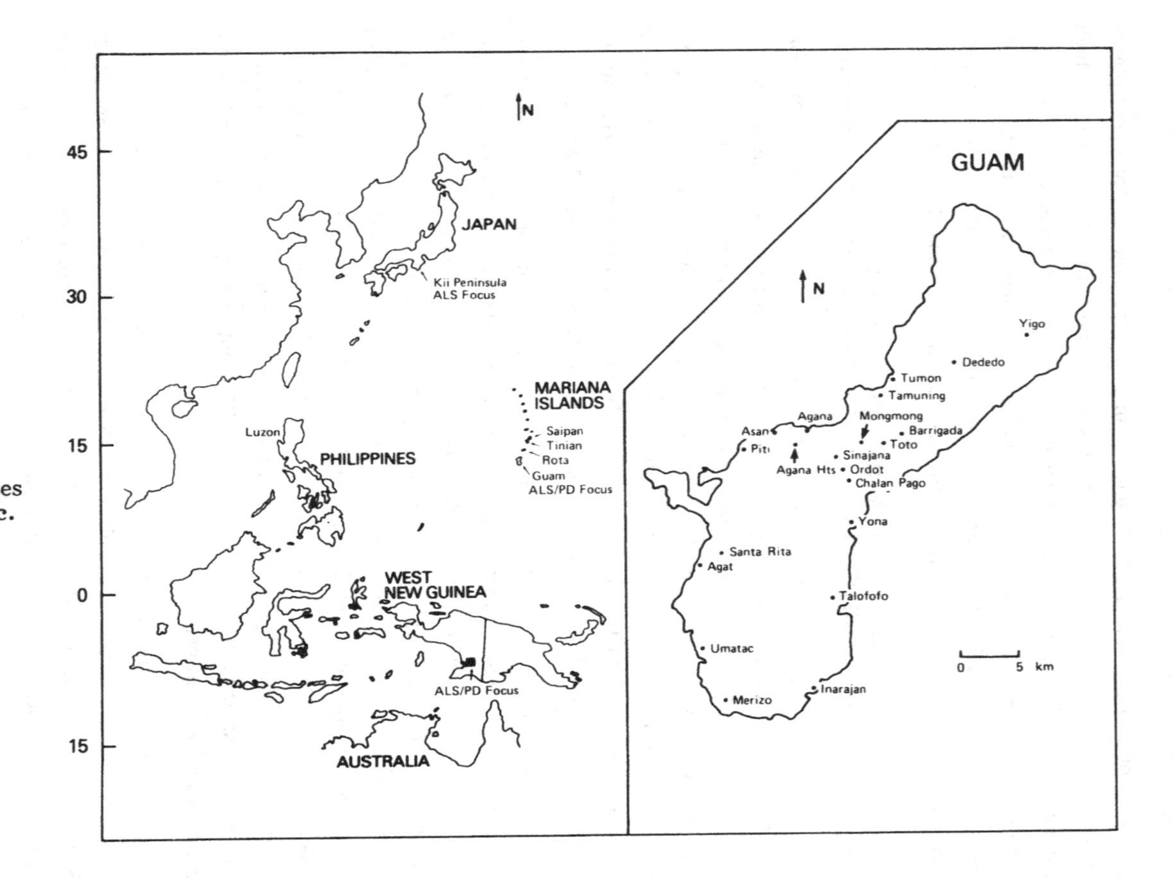

Figure 2:

Location of the three high incidence foci of amyotrophic lateral sclerosis and parkinsonian syndromes in the western Pacific.

that of the U.S. mainland and the highest recorded level in the world. In selected villages, the prevalence exceeds 1300 per 100,000 population. The focus is geographically discrete, and repeated surveys have established that neither ALS nor parkinsonism is found outside the focal region in the surrounding larger population.

Of the three known foci of ALS and parkinsonism, none has been more closely monitored epidemiologically, clinically and neuropathologically than among the Chamorros of Guam and the Northern Mariana Islands (Garruto, 1985; Hirano et al, 1961a, 1961b; Kurland & Mulder, 1954; Rogers-Johnson et al, 1986). The striking concentration of ALS in the Mariana Islands led to the establishment of a field station on Guam in 1956 by the National Institute of Neurological and Communicative Disorders and Stroke (then known as the National Institute of Neurological Diseases and Blindness). The primary function of the research program on Guam has been the epidemiological, clinical and neuropathological surveillance of these disorders in combination with specialised studies concerned with identifying etiological factors and mechanisms of pathobiology. A brief summary of this focus is presented here. Detailed reviews of ALS and PD in the western Pacific have been presented elsewhere (Gajdusek, 1984; Garruto & Gajdusek, 1984).

THE GUAMANIAN FOCUS: A SYSTEMATIC SEARCH FOR CAUSAL MECHANISMS OF DISEASE

ALS, locally referred to on Guam as *lytico*, had an original incidence more than 50 times that of the continental United States (Brody & Kurland, 1973). Subsequent research revealed the presence of a second neurological disorder, parkinsonism-dementia (PD), locally referred to as *bodig*, which also occurs at a high incidence among the Chamorro people and is found in close association with the high incidence of ALS. Both disorders are chronically progressive and uniformly fatal, and previously accounted for 1 in 5 deaths among adult Guamanians over age 25. Although the exact relationship between these disorders in each of the three foci is unknown, they may represent different disease expressions resulting from a common etiologic pathway.

Historically, the original ancestors of the native Chamorros are presumed to have migrated from Malaysia and the Philippine Islands as early as 2000 BC. Europeans first arrived on Guam with Magellan in 1521. Under early Spanish rule, the population declined dramatically as the result of the Spanish-Chamorro Wars, severe typhoons, and disease. The declining native Chamorro population intermarried extensively in the late 1700's, particularly with Filipinos and, to a much lesser extent, with Mexican and Spanish colonials (Thompson, 1947). Until recently, the population has remained genetically stable, although its extensive early admixture modified the original gene pool such that "pure" Chamorros are unlikely to exist today.

A major consideration in assessing risk factors for ALS and PD has been the relative contribution of genetics and environment to these disorders. The kinds of genetic analyses that have been systematically performed on Guam are summarised in Figure 3. These extensive studies initially suggested an autosomal dominant disease with variable or incomplete penetrance. There is clearly a familial and sporadic form of disease but no clear genetic pattern has, as yet, been identified. In addition to a Mendelian dominant theory with incomplete penetrance, hypotheses including a polygenetic form of inheritance, a genetic susceptibility, or a multifactorial threshold effect have been offered. Although

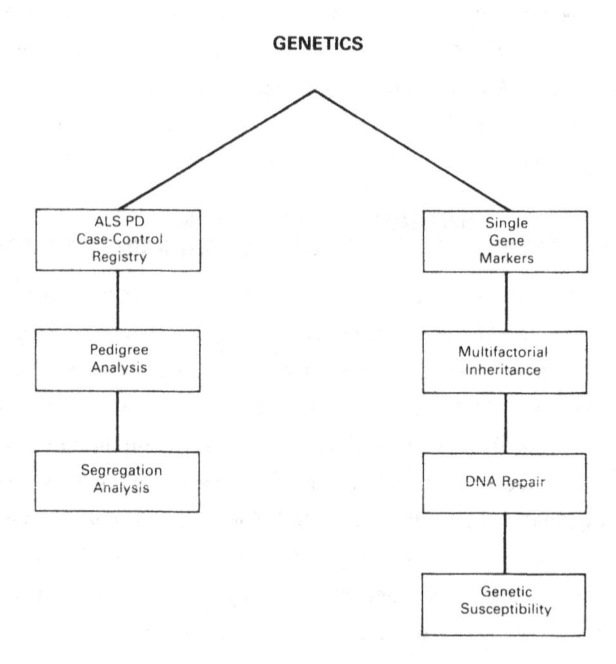

Figure 3: Summary of genetic studies conducted in the Guamanian focus of amyotrophic lateral sclerosis (ALS) and parkinsonism-dementia (PD).

both ALS and PD are known to occur together in the same Chamorro family, the same sibship, and even occasionally in the same individual, intensive genetic studies including pedigree analysis of high incidence villages, calculation of inbreeding coefficients for such villages, the development of prospective case-control registries, and the search for specific gene markers, such as HLA antigens at the A and B locus, blood group systems, red cell enzymes, immunoglobulin allotypes, serum proteins, and dermatoglyphics, have not yielded a satisfactory genetic explanation (Blake et al, 1983; Garruto et al, 1983; Hoffman et al, 1977; Plato et al, 1967; Plato et al, 1969; Reed et al, 1975). However, a genetic susceptibility in combination with an "environmental trigger" remains a distinct possibility.

Systematic epidemiological observations over three decades have clearly been instrumental in the determination of potential causal factors associated with ALS and PD on Guam. These studies have included 30 years of epidemiological assessment in a comparatively small population allowing nearly complete case ascertainment of these diseases. Descriptive and analytic epidemiological studies have included the assessment of both native Chamorros and non-Chamorro migrants to Guam, the surveillance of ALS and PD in the Mariana Islands north of Guam and an evaluation of these diseases in Chamorro migrants to the continental United States (Figure 4). Such studies could not have been conducted in large western cosmopolitan communities to the extent they have on Guam.

During the past 30 years, the incidence and mortality rates of ALS and PD have dramatically declined and today the risk to Guamanian Chamorros is only a several times that of non-Chamorro residents on the United States mainland (Garruto et al, 1985). This decline and related epidemiological data strongly incriminate environmental factors as a primary cause of disease. The previously high incidence of ALS and PD three decades ago is consistent with a cohort phenomenon involving exposure to selected cultural/environmental factors; both diseases nearly disappeared during the aggressive acculturation from a subsistence economy to today's almost completely westernized culture with a cash economy. Similar declines are evident in the Kii Peninsula focus of Japan (Uebayashi, 1980) and more recently in two villages in the West New Guinea focus where western contact and introduction of new foodstuffs has occurred (Gajdusek, 1984).

Efforts to transmit ALS and PD experimentally to non-human primates and other laboratory animals have been unsuccessful (Gibbs & Gajdusek, 1982), as

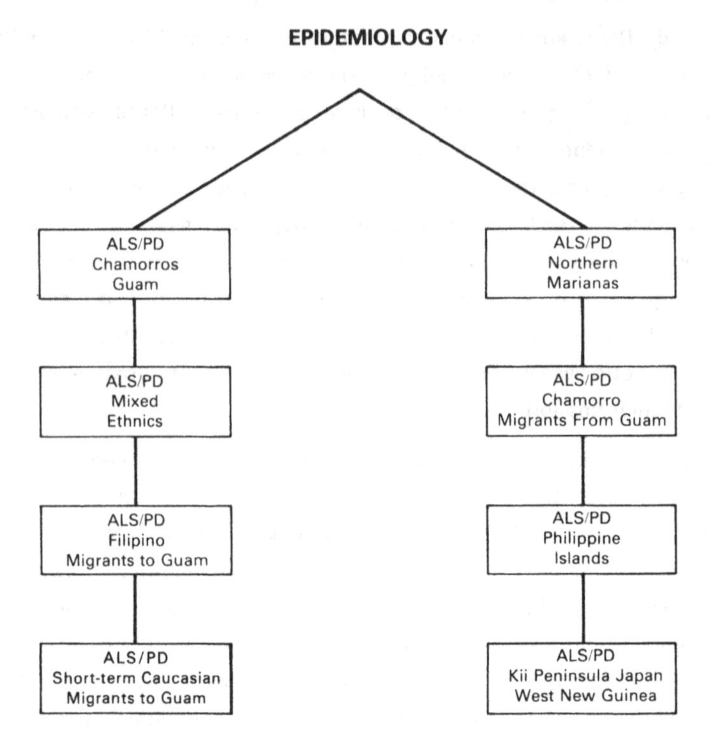

Figure 4: Summary of epidemiological investigations of amyotrophic lateral sclerosis (ALS) and parkinsonian syndromes (PD) in the western Pacific during the past 30 years of surveillance.

has the search for a virus by cultivation of brain tissue *in vitro*, by inoculation of cell cultures, or by attempts to find a viral protein, antigen or nucleic acid in the tissues of patients (Kohne et al, 1981; Viola et al, 1979) (Figure 5). Nor has an association between ALS and PD and plant and animal toxins become evident thus far, despite extensive research on the cycad nut, cassava, and fish toxins on Guam (Kurland, 1978; Reed et al, 1966).

However deficiencies and excesses of certain trace and essential elements in soil and water are consistently found in all three foci (Garruto et al, 1984b). It has been postulated that abnormalities in the mineral environment are involved in ALS and PD (Gajdusek, 1985; Garruto, 1985; Yase, 1972). Chronic nutritional deficiencies of calcium may lead to abnormalities in mineral metabolism, which, in the presence of excessive accumulation of certain divalent and trivalent cations, result in the deposition of these elements in neurons

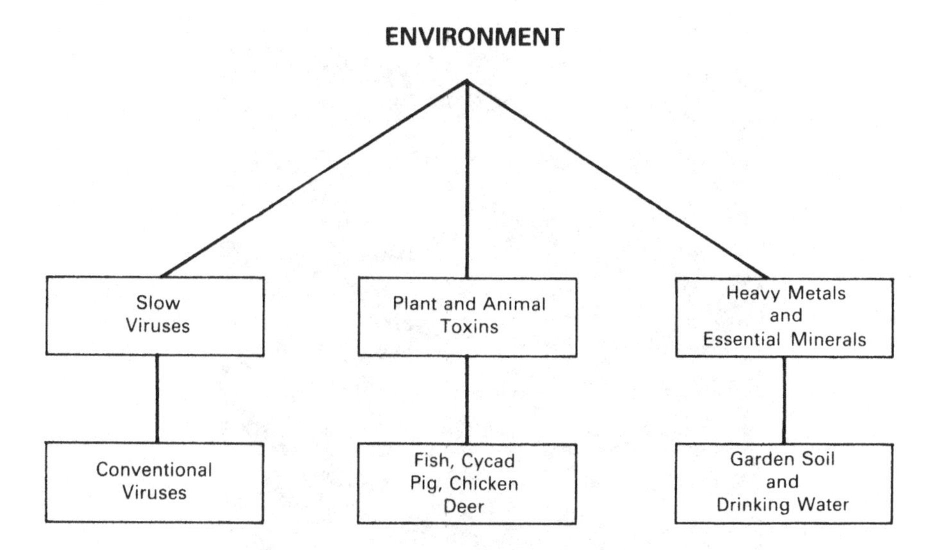

Figure 5: Summary of environmental investigations in high incidence western Pacific foci of amyotrophic lateral sclerosis and parkinsonian syndromes.

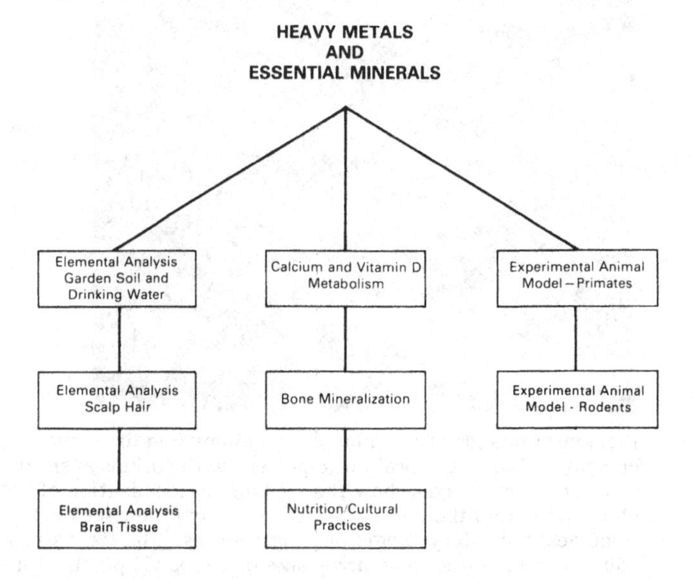

Figure 6: Summary of attempts to identify an association between heavy metals and essential minerals in high incidence western Pacific foci of amyotrophic lateral sclerosis and parkinsonian syndromes.

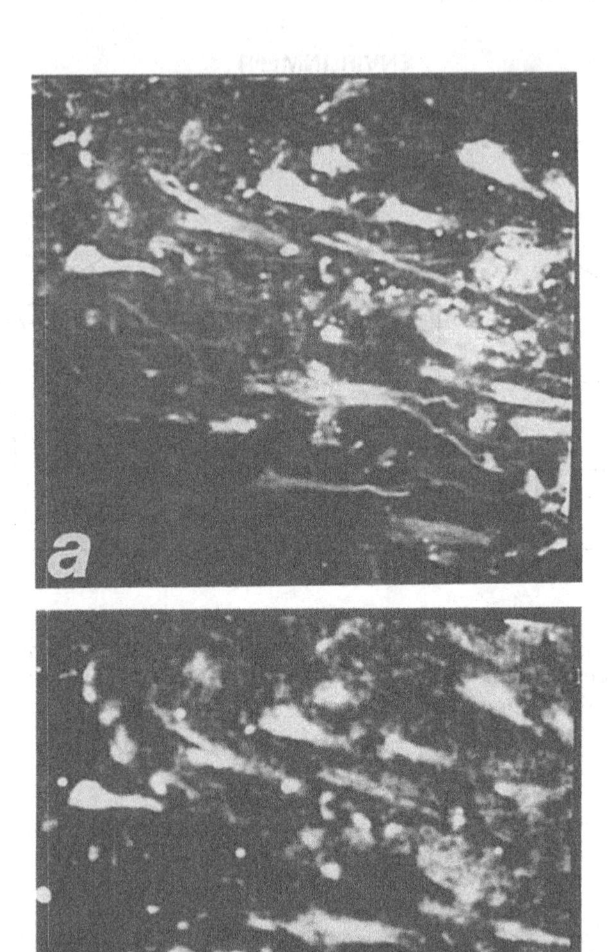

Figure 7: Elemental images of calcium (a) and aluminium (b) in the hippocampal region of the brain of a patient with parkinsonism-dementia of Guam. The images show the striking co-localisation of both elements within the cell body and dendritic processes of the same neurofibrillary-tangle bearing neurons. The field size is 200 µm x 200 µm with an array size of 256 x 256 pixels. The image was taken using a new method of computer-controlled electron-beam x-ray microanalysis (Garruto et al, 1984a).

possibly as hydroxyapatites, culminating eventually in cell death. The involvement of chronic deficiencies and excesses of alkaline earth metals in the pathogenesis of ALS and PD has recently been supported by the reported disturbances in calcium and vitamin D metabolism in patients, cortical bone loss in Guamanian children and adults, and the intraneuronal accumulation of calcium, aluminium, and silicon in brain tissue from affected patients (Garruto et al, 1984a; Plato et al, 1982, 1984; Yanagihara, 1982; Yanagihara et al, 1984; Yase, 1980; Yoshimasu et al, 1982) (Figures 6, 7).

CONCLUSION

The intriguing medical enigma of ALS and associated parkinsonism and dementia, and perhaps the aging process *per se* continues to unfold as these unique hyperendemic foci are studied. Such isolated Pacific foci represent natural experimental models for the elucidation of risk factors and mechanisms for the origin and dissemination of disease to an extent not possible in larger, westernised societies. The natural decline of ALS and parkinsonism on Guam and the Kii Peninsula continues, thus alleviating more than a century of burden for populations which have suffered inordinately from this effect.

Our ability to gain insights into the etiology and pathogenesis of major neurological disorders of widespread medical significance has been enhanced by the use of isolates as natural experimental phenomena. We must not, however, attempt to arrive at some parochial scheme of research with regard to isolates but, rather, to capitalise on the unique features of each to answer the particular problems for which it is most suited. Consequently, the study of disease patterns in isolated groups should be holistic, integrative, and necessarily opportunistic.

REFERENCES

Blake, N. M., Kirk, R. L., Wilson, S. R., Garruto, R. M., Gajdusek, D. C., Gibbs, C. J. Jr. & Hoffman, P. (1983). Search for a red cell enzyme or serum protein marker in amyotropic lateral sclerosis and parkinsonism-dementia of Guam. Americal Journal of Medical Genetics, **14**, 299-305.

Brody, J. A. & Kurland, L. T. (1973). Amyotrophic lateral sclerosis and parkinsonism-dementia in Guam. In: J.D. Spillane (ed.), Tropical Neurology, pp. 355-375. London: Oxford University Press.

Gajdusek, D. C. (1963). Motor-neuron disease in natives of New Guinea. The New England Journal of Medicine, **268**, 474-476.

Gajdusek, D. C. (1984). Environmental factors provoking physiological changes which induce motor neuron disease and early neuronal aging in high incidence foci in the western Pacific: calcium deficiency-induced secondary hyperparathyroidism and resultant CNS deposition of calcium and other metallic cations as the cause of ALS and PD in high incidence foci. In: F.C. Rose (ed.), Progress in Motor Neuron Disease, pp. 44-69. Kent: Pitman Books.

Gajdusek, D. C. (1985). Hypothesis: Interference with axonal transport of neurofilament as a common pathogenetic mechanism in certain diseases of the central nervous system. The New England Journal of Medicine, **312**, 714-719.

Gajdusek, D. C. & Salazar, A. M. (1982). Amyotrophic lateral sclerosis and parkinsonian syndromes in high incidence among the Auyu and Jakai people of West New Guinea. Neurology, **32**, 107-126.

Garruto, R. M. (1985). Elemental insults provoking neuronal degeneration: the suspected etiology of high incidence amyotrophic lateral sclerosis and parkinsonism-dementia of Guam. In: J.T. Hutton & A.D. Kenny (eds.), Senile Dementia of the Alzheimer Type, pp. 319-336. New York: Alan R. Liss.

Garruto, R. M. & Gajdusek, D. C. (1984). Pacific cultures: a paradigm for the study of late onset neurological disorders. In: H. Rothschild (ed.), Risk Factors for Senility, pp. 74-89. New York: Oxford University Press.

Garruto, R. M., Plato, C. C., Myrianthopoulos, N. C., Schanfield, M. S. & Gajdusek, D. C. (1983). Blood groups, immunoglobulin allotypes and dermatoglyphic features of patients with amyotrophic lateral sclerosis and parkinsonism-dementia of Guam. American Journal of Medical Genetics, **14**, 289-298.

Garruto, R. M., Fukatsu, R., Yanagihara, R., Gajdusek, D. C., Hook, G. & Fiori, C. E. (1984a). Imaging of calcium and aluminium in neurofibrillary tangle-bearing neurons in parkinsonism-dementia of Guam. Proceedings of the National Academy of Sciences, **81**, 1875-1879.

Garruto, R. M., Yanagihara, R., Gajdusek, D. C. & Arion, D. M. (1984b). Concentrations of heavy metals and essential minerals in garden soil and drinking water in the western Pacific. In: K.-M. Chen & Y. Yase (eds.), Amyotrophic Lateral Sclerosis in Asia and Oceania, pp. 265-329. Taipei: National Taiwan University.

Garruto, R. M., Yanagihara, R. & Gajdusek, D. C. (1985). Disappearance of high incidence amyotrophic lateral sclerosis and parkinsonism-dementia on Guam. Neurology, **35**, 193-198.

Gibbs, C. J. Jr. & Gajdusek, D. C. (1982). An update on long-term in vivo and in vitro studies designed to identify a virus as a cause of amyotrophic lateral sclerosis, parkinsonism-dementia and Parkinson disease. In: L.P. Rowland (ed.), Advances in Neurology, Vol. 3, Human Motor Neuron Diseases, pp. 343-353. New York: Raven Press.

Hirano, A., Kurland, L. T., Krooth, R. S. & Lessell, S. (1961). Parkinsonism-dementia complex, an endemic disease on the island of Guam. I. Clinical features. Brain, **84**, 642-661.

Hirano, A., Malamud, N. and Kurland, L. T. (1961). Parkinsonism-dementia complex, an endemic disease on the island of Guam. II. Pathological features. Brain, **84**, 662-679.

Hoffman, P. M., Robbins, D. S., Gibbs, C. J. Jr., Gajdusek, D. C., Garruto, R. M. & Terasaki, P. I. (1977). Histocompatibilty antigens in amyotrophic lateral sclerosis and parkinsonism-dementia on Guam. Lancet, **2**, 717.

Kohne, D. E., Gibbs, C. J. Jr., White, L., Tracy, S. M., Meinke, W. & Smith, R. A. (1981). Virus detection by nucleic acid and hybridization: examination of normal and ALS tissues for the presence of poliovirus. Journal of General Virology, **56**, 223.

Kurland, L. T. (1978). Geographic isolates: their role in neuroepidemiology. In: B.S. Schoenberg (ed.), Advances in Neurology, Vol. 19, Neurological Epidemiology: Principles and Clinical Applications, pp. 69-82. New York: Raven Press.

Kurland, L. T. & Mulder, D. W. (1954). Epidemiologic investigations of amyotrophic lateral sclerosis. I. Preliminary report on geographic distribution, with special reference to the Mariana Islands, including clinical and pathological observations. Neurology, **4**, 355-378, 438-448.

Plato, C. C., Reed, D. M., Elizan, T. S. & Kurland, L. T. (1967). Amyotrophic lateral sclerosis/parkinsonism-dementia complex of Guam. IV. Familial and genetic investigations. American Journal of Human Genetics, **19**, 617-632.

Plato, C. C., Cruz, M. T. & Kurland, L. T. (1969). Amyotrophic lateral sclerosis/parkinsonism-dementia complex of Guam: further genetic investigations. American Journal of Human Genetics, **21**, 133-141.

Plato, C. C., Garruto, R. M., Yanagihara, R., Chen, K.-M., Wood, J. L., Gajdusek, D. C. & Norris, A. H. (1982). Cortical bone loss and measurements of the second metacarpal bone. I. Comparisons between adult Guamanian Chamorros and American Caucasians. American Journal of Physical Anthropology, **59**, 461-465.

Plato, C. C., Greulich, W. W., Garruto, R. M. & Yanagihara, R. (1984). Cortical bone loss and measurements of the second metacarpal. II. Hypodense bone in post-war Guamanian children. American Journal of Physical Anthropology, **63**, 57-63.

Reed, D. M. & Brody, J. A. (1975). Amyotrophic lateral sclerosis and parkinsonism-dementia on Guam, 1945-1972. I. Descriptive epidemiology. American Journal of Epidemiology, **101**, 287-301.

Reed, D., Plato, C., Elizan, T. & Kurland, L. T. (1966). The amyotrophic lateral sclerosis/parkinsonism-dementia complex: a ten-year follow-up on Guam. Part I. Epidemiologic studies. American Journal of Epidemiology, **83**, 54.

Reed, D. M., Torres, J. M. & Brody, J. A. (1975). Amyotrophic lateral sclerosis and parkinsonism-dementia on Guam, 1945-1972. II: Familial and genetic studies. American Journal of Epidemiology, **101**, 302-310.

Rogers-Johnson, P., Garruto, R. M., Yanagihara, R., Chen, K-M., Gajdusek, D. C. & Gibbs, C. R. Jr. (1986). Guamaniam amyotrophic lateral sclerosis and parkinsonism-dementia on Guam: a 30-year evaluation on clinical and neuropathological trends. Neurology, **36**, 7-13.

Shiraki, H. & Yase, Y. (1975). Amyotrophic lateral sclerosis in Japan. In: G.W. Bruyn (ed.), Handbook of Clinical Neurology, pp. 353-419. New York: Elsevier.

Thompson, L. (1947). Guam and Its People. Princeton: Princeton University Press.

Uebayashi, Y. (1980). Epidemiological investigation of motor neuron disease in the Kii Peninsula, Japan, and on Guam - the significance of long survival cases. Wakayama Medical Reports, **23**, 13-27.

Viola, M. V., Myers, J. C., Gann, K. L., Gibbs, C. J. Jr. & Roos, R. P. (1979). Failure to detect poliovirus genetic information in amyotrophic lateral sclerosis. Annals of Neurology, **5**, 402.

Yanagihara, R. (1982). Heavy metals and essential minerals in motor neuron disease. In: L.P. Rowland (ed.), Human Motor Neuron Diseases, pp. 233-247. New York: Raven Press.

Yanagihara, R., Garruto, R. M., Gajdusek, D. C., Tomita, A., Uchikawa, T., Konagaya, Y., Chen, K-M., Sobue, I., Plato, C. C. & Gibbs, C. J. Jr. (1984). Calcium and vitamin D metabolism in Guamanian Chamorros with amyotrophic lateral sclerosis and parkinsonism-dementia. Annals of Neurology, **15**, 42-48.

Yase, Y. (1972). The pathogenesis of amyotrophic lateral sclerosis. Lancet, **2**, 292-296.

Yase, Y., (1980). The role of aluminium in CNS degeneration with interaction of calcium. Neurotoxicology, **1**, 101-109.

Yoshimasu, F., Yasui, M., Yase, Y., Uebayashi, Y., Tanaka, S., Iwata, S., Sasajima, K., Gajdusek, D. C., Gibbs, C. J. Jr. & Chen, K-M. (1982). Studies on amyotrophic lateral sclerosis by neutron activation analysis. II. Systemic analysis of metals on Guamanian ALS and PD cases. Folia Psychiatrica et Neurologia Japonica, **36**, 173-179.

INDEX